John Abercrombie

The Student's Guide to Medical Jurisprudence

John Abercrombie

The Student's Guide to Medical Jurisprudence

ISBN/EAN: 9783337312831

Printed in Europe, USA, Canada, Australia, Japan

Cover: Foto ©berggeist007 / pixelio.de

More available books at **www.hansebooks.com**

THE

STUDENT'S GUIDE

TO

MEDICAL JURISPRUDENCE

BY

JOHN ABERCROMBIE, M.D. Cantab., M.R.C.P.

LECTURER ON FORENSIC MEDICINE AT THE CHARING CROSS HOSPITAL
MEDICAL SCHOOL, AND ASSISTANT-PHYSICIAN TO THE HOSPITAL
CORRESPONDING MEMBER OF THE MEDICO-LEGAL SOCIETY OF NEW YORK

LONDON
J. & A. CHURCHILL
11 NEW BURLINGTON STREET
1885

PREFACE.

THIS little book has been compiled for the use of students at the suggestion of the publishers. The freest use has been made of the standard works of Taylor, Tidy, Wharton and Stillé, Casper (New Sydenham Society's translation), Ogston, Woodman and Tidy, and Guy and Ferrier, as also of Taylor on poisons and Winter Blyth on poisons. To those who have the means and leisure to study any of the above, such a work as this is not addressed, but it is believed that there are many within whose reach they never come. In dealing with the poisons the writer has purposely avoided all mention of the search for them in organic compounds, partly because such knowledge can only be really acquired in the laboratory, and partly because in every-day practice analyses of this nature are invariably conducted by an analyst who has devoted special attention to this particular subject. The writer can lay but little claim to originality, he has simply endeavoured to condense and group together, in a readily assimilable form, the chief facts belonging to the important subject of Medical Jurisprudence.

April 15, 1885.

CONTENTS.

INTRODUCTION.
DEFINITION.

The medical man in the witness-box—Expert evidence—Dying declarations—Post-mortem examinations . . . . PAGE 1

CHAPTER I.
THE EARLY SIGNS OF DEATH.

Cessation of circulation and respiration—Insensibility—Loss of temperature—State of the skin—Muscular flaccidity—Rigor mortis—Cadaveric spasm 8

CHAPTER II.
THE LATER SIGNS OF DEATH.

Putrefaction—Modifying conditions—Putrefaction in the earth—Adipocere—Putrefaction in water—Mummification . 23

CHAPTER III.
IDENTITY.

Mutilated remains—The age of a skeleton—Identity from the teeth—Determination of the height—Determination of the sex—The female pelvis 37

CHAPTER IV.
SUDDEN DEATH AND PRESUMPTION OF DEATH.

Death commencing at the heart, brain, and lungs—Pretended death—Presumption of survivorship 52

CHAPTER V.
DEATH FROM COLD, HEAT, AND STARVATION 61

CHAPTER VI.
DEATH FROM ASPHYXIA.
Definition — Symptoms — Post-mortem appearances — Death from drowning—Mode of death—Treatment—Post-mortem appearances—Cause of death 69

CHAPTER VII.
DEATH FROM ASPHYXIA (*continued*).
Death from hanging—Judicial hanging—Symptoms—Treatment—The mark on the neck—Death from strangulation—Death from suffocation 88

CHAPTER VIII.
DEATH FROM LIGHTNING 99

CHAPTER IX.
WOUNDS.
Definition—Classification—Ante- and post-mortem wounds—Incised, contused, lacerated, and punctured wounds . . 102

CHAPTER X.
CAUSATION OF WOUNDS, ETC.
Suicide or murder—Character of the wound—Direction of the wound—Blood stains—Their examination 109

CHAPTER XI.
THE CAUSE OF DEATH FROM WOUNDS.
Hæmorrhage—Shock—Mechanical injury—Wounds indirectly fatal—Death caused by the treatment—Self-inflicted wounds 122

CHAPTER XII.
WOUNDS OF THE DIFFERENT REGIONS OF THE BODY.
Head—Brain—Face—Neck—Vertebræ—Railway injuries . 132

CHAPTER XX.

IRRITANT POISONS (*continued*).

Caustic potash and soda—Ammonia—Nitre—Chlorate of potash—Sulphate of potash—Alum—Salts of Barium—Oxalic acid—Tartaric and acetic acids 201

CHAPTER XXI.

IRRITANT POISONS (*continued*).

Phosphorus—Iodine—Bromine—Arsenic—Antimony . . 214

CHAPTER XXII.

IRRITANT POISONS (*continued*).

Copper—Mercury—Lead—Zinc—Iron—Silver—Bichromate of potash 232

CHAPTER XXIII.

VEGETABLE AND ANIMAL IRRITANTS.

Savin—Croton oil—Hellebore—Colchicum—Ergot—Cantharides—Poisonous foods—Ptomaines 246

CHAPTER XXIV.

NEUROTIC POISONS.

Opium—Differential diagnosis—Opium-eating—Belladonna—Hyoscyamus—Stramonium—Camphor—Nitro-benzine and Aniline—Poisonous fungi 252

CHAPTER XXV.

NEUROTIC POISONS (*continued*).

Alcohol—Ether—Chloroform—Hydrate of Chloral—Bisulphide of carbon—Carbolic acid 265

CHAPTER XXVI.

NEUROTIC POISONS (*continued*).

Strychnia — Conium — Œnanthe crocata — Cicuta virosa — Æthusa cynapium—Curare—Taxus baccata—Physostigma faba—Tobacco—Cocculus indicus—Lobelia . . . 276

CHAPTER XXVII.
NEUROTIC POISONS (*continued*).

Prussic acid—Turpentine—Nitro-glycerine—Digitalis—Aconite 290

CHAPTER XXVIII.
POISONOUS GASES.

Carbonic acid—Carbonic oxide—Coal gas—Sulphuretted hydrogen 302

CHAPTER XXIX.
PREGNANCY.

General symptoms—Local signs—Signs referable to the fœtus—Feigned and concealed pregnancy—Unconscious pregnancy 305

CHAPTER XXX.
DELIVERY.

Signs of recent delivery in the living—In the dead—The corpus luteum—Signs of delivery at a remote period—Characters of the embryo—Concealment of birth 314

CHAPTER XXXI.
ABORTION.

Definition—Statutes relating to—Causes: Natural—Violent—Savin—Ergot—Signs of abortion 324

CHAPTER XXXII.
LEGITIMACY—DURATION OF PREGNANCY.

Viability—Maturity—Superfœtation—Monsters—Disputed Paternity—Hermaphroditism 33

CHAPTER XXXIII.

IMPOTENCE, STERILITY, RAPE, UNNATURAL OFFENCES . . 346

CHAPTER XIII.

WOUNDS OF THE DIFFERENT REGIONS OF THE BODY (*continued*).

Chest—Heart—Abdomen—Genital organs—Fractures and dislocations 139

CHAPTER XIV.

GUNSHOT WOUNDS, CICATRICES, BURNS AND SCALDS.

Wounds of entrance and of exit—The weapon—Cicatrices—Tattooing—Scars from disease—Burns and scalds—Zone of redness—Formation of blisters—Spontaneous combustion . 144

CHAPTER XV.

LIFE INSURANCE.

Sound lives—Unsound lives—Material concealment—Intemperance—Suicide 154

CHAPTER XVI.

Vision 165

CHAPTER XVII.

INSANITY.

Definition—Classification—Contracts—Wills—State control of lunatics—Certificates of Lunacy—Interdiction—The plea of insanity—Epileptic mania—Feigned insanity—Intoxication 167

CHAPTER XVIII.

POISONS.

Statutes relating to—Definition—Mode of action—Causes modifying the action—General evidence of poisoning . . 183

CHAPTER XIX.

IRRITANT POISONS.

The mineral acids—Sulphuric acid—Symptoms—Post-mortem appearances—Treatment—Tests—Nitric acid—Hydrochloric acid 190

CHAPTER XXXIV.

BIRTH AND INHERITANCE.

Live birth—Inheritance—Legal relationships of age . . 356

CHAPTER XXXV.

INFANTICIDE.

Signs of maturity—Intra-uterine maceration—Signs of life after birth—Tests of live birth—Causes of death . 361

INDEX 383

STUDENT'S GUIDE

TO

MEDICAL JURISPRUDENCE.

INTRODUCTION.

MEDICAL jurisprudence, or forensic medicine, has been aptly defined to be 'that science which teaches the application of every branch of medical knowledge to the purposes of the law.' If ever there was a time when the separate study of medical jurisprudence was unnecessary, it is quite certain that such a time has gone by. Not a day passes without ample evidence of this fact being afforded in the columns of the newspapers. Either an old man's will is disputed on the ground that his mind was affected at the time it was made; or the dead body of an infant is found in some out-of-the-way place, and the coroner's jury have to decide whether the child had been born alive, and if so, what was the cause of death; or the plea of insanity is set up by the defence in a case of murder; or a question as to the identity of a dead body is raised; or a claim to an estate is resisted on the ground of the illegitimacy of the claimant. Many other instances might be quoted as of daily occurrence, all of them cases in which medical evidence is required before the point at issue can be determined.

A good knowledge of medicine, surgery, obstetrics, anatomy, physiology, chemistry, physics, and botany will doubtless go far towards making a man a good medical jurist; but in themselves they are insufficient, and unless he has studied the rules of evidence, and trained himself to formulate his ideas definitely and to be able to support his conclusions by clearly-expressed reasons, the medical witness may find that all his learning will stand him in very poor stead in the witness-box.

In all departments of medicine a certain degree of scepticism about the statements of a patient's friends is wholesome. In medico-legal questions, such scepticism is more than ever needed. The medical man should listen to the information that the friends may have to impart, and receive it courteously; but if it does not tally with the results of his own observation, he must not place too much reliance upon it, remembering that he never can tell what motives are at work.

The subjects which have to be considered under the head of Medical Jurisprudence are the questions relating to the dead body, such as the age, sex, date of death, cause of death; the various modes of death from violence; injuries from whatever cause; life assurance; insanity; and the whole subject of poisoning. Certain questions in regard to the pregnant state and the duration of pregnancy, impotence, sterility and rape will also have to be considered, as also the important subject of infanticide or child-murder.

The medical man in the witness-box. Before proceeding to deal with these subjects separately, a few words may be said about evidence in court, whether before the coroner, magistrate, or judge and jury. Every legally-qualified medical man is liable any day to find himself in the witness-box, and it is therefore essential that he should have paid some attention to the subject of evidence, not only for his own sake, but also for the credit of the whole profession. A medical man may be put into the witness-box in one of two capacities. Either to testify to facts, as for instance if he is summoned to give evidence as to the discovery of a dead body, and as to the exact state in which it was when he was called to it—when subpœnaed to give evidence under such circumstances a medical man has no option, but must attend the summons; or he may be called to express his scientific opinion on a case merely from the evidence given in court—no man is bound to accept a subpœna of this kind. In any case, before entering the witness-box the medical man should carefully think over the facts of the case, and be quite clear in his own mind as to the opinion he has formed and the reasons for that opinion. An American writer, alluding to this subject, has said: 'A man, whether learned or not, will talk clearly upon a subject he well understands, whether it is scientific or not; but unless it is clear in his own mind, his account of it will be confused and unsatisfactory.' The witness must always use the simplest language he can command to express his meaning, and should scrupulously shun the use of technical terms, as also of an exaggerated style of speaking. Before answering a question he should be sure that he understands it, and if he is asked any question he cannot answer from want of knowledge he should at once say so. Unless his reply would tend to incriminate himself, a witness is bound to answer any question that is put to him, or run the risk of being committed for contempt of court.

An expert, when called, must not quote what others have written; he is called to give his own opinions only, and not those of others. As a general principle, it is contrary to custom that the works of any living author should be quoted, on the ground that, if that author's opinion is required, he

Expert evidence.

should be called in court and give his opinion on oath. If the witness is repeating some remarks made to him by the prisoner or in his presence, he must quote the actual words used, and not simply paraphrase the gist of the conversation.

The use of notes. In regard to the use of notes at a trial, a witness is not allowed to read his notes, but he may refer to them to refresh his memory if they were notes actually made at the time of the occurrence to which they allude. Thus, in the case of a post-mortem examination, the witness may only use the notes actually made at the time, which must not have been subsequently amended or altered.

Dying declarations. When a person is dying, it may be that he wishes to make some statement of great importance, either as to the cause of his death or in reference to the disposal of his property. In such a case a magistrate ought to be called in to receive the deposition of the dying man. In cases of death from violence, it will often happen that there is not time to send for a magistrate, and then the duty of receiving the declaration of the dying man usually devolves on the medical man. In these circumstances he should, if possible, get the man to write down on paper the statement he wishes to make and sign it, the medical man and any others present attesting its accuracy by adding their signatures. Should the man be too ill, the medical man should write it himself, taking care only to write such words as are dictated to him, and not to make suggestions. After it is finished he should read it through to the dying man, and get him, if possible, to sign it, and then ask any of those present to witness it. A document of this character will always be allowed to be put in evidence in court. The only grounds on which it could be disputed would be, either that the man making the declaration did not make it in the belief that he was dying, which is always an essential point, or that the man was not in his right mind and did not understand the nature of his acts. On both these points the medical man will be able to give evidence that ought to be conclusive.

It cannot be too strongly impressed upon students that they ought to lose no opportunity of making themselves familiar with the appearances of the viscera in the different diseases. In a trial for murder by poisoning, the defence will generally be that the deceased died from natural causes; and unless a man has a tolerably thorough knowledge of gross morbid anatomy, he will not be able to disprove this theory.

Every medical man is supposed to know a fatty heart when he sees one, or to be able to say whether there was any cirrhosis of the liver, or whether the kidneys were healthy or not; but, unless during his student-days he has been a diligent attendant in the post-mortem room, he will certainly not be able to answer such questions with any degree of confidence.

When making a post-mortem examination for medico-legal purposes, the practitioner should first of all note the condition of the body, and if it has not been laid out in the usual way, he should observe the attitude of the limbs. He should then note whether rigor mortis is present or not, and the degree of warmth (if any) that is present; these points would assist him in forming an opinion as to the time that had elapsed since death. *Post-mortem examination, how performed.*

The body should next be carefully examined for external marks of violence, such as bruises or wounds, care being taken not to confound post-mortem lividities with bruises inflicted during life. The differences between these will be discussed in the text. Wounds should be carefully examined, their size, shape, position and direction being noted, and they should be probed to see if any bullet or other foreign body can be reached. A consideration of these various points will assist the practitioner in arriving at a conclusion as to the weapon with which they were probably inflicted.

The internal viscera should then be examined; in the case of bullet wounds or punctured wounds of any sort, the practitioner should trace the exact course of the bullet or penetrating instrument. In all cases, even when the cause of death has been ascertained beyond doubt, it would be wise to make a complete examination, in order to be able to deny the possibility of any other cause of death. The mouth, pharynx and larynx should be examined, as sometimes persons have died suddenly choked by food or false teeth passing into the larynx, and if this precaution were neglected the cause of death might remain obscure. In all cases where there is the slightest likelihood of the question of poisoning being raised, a ligature should be placed at the lower end of the œsophagus, and a double ligature on the duodenum, the stomach can then be removed with all its contents. If another ligature be placed on the rectum, the intestines can be removed in a similar manner. These may then be opened over separate clean jars, so that the contents are saved and the mucous membrane of the whole alimentary canal can be examined. After this has been done, the stomach should be placed in a clean jar and also the intestines; each jar being hermetically sealed at the time and placed under lock and key. Some of the urine should be saved, as also portions of the liver, spleen and kidneys, and similarly taken care of. These will ultimately be handed over to the analyst appointed to examine them, but until they are placed in his charge it is essential that the practitioner should have so kept them that it is impossible that they should have been tampered with. The practitioner should make notes of the condition of the various internal viscera, and of the exterior of the body, whilst they are before his eyes. From these notes he can subsequently draw up his report, should he be asked to send one in.

CHAPTER I.

THE EARLY SIGNS OF DEATH.

<small>Importance of the immediate signs of death.</small>

THE importance of being able to determine with accuracy whether death has taken place or not, will be obvious to all. It is necessary therefore to examine somewhat carefully the signs which have been propounded as evidence of death. These may be divided into two groups, viz. those which are due to post-mortem changes, and those which may be looked for before sufficient time has elapsed for the commencement of such changes. This latter group, the immediate signs of death as they have been called, has naturally a much greater value than the others, because it is clearly desirable that when a person is dead the fact should be recognised as soon as possible. The immediate signs of death can only be satisfactorily determined by a medical man, and very little reliance should be placed on the opinion of a layman as to the fact of the cessation of the circulation, whereas the advent of rigor mortis or the supervention of putrefaction could be as easily recognised by a layman as by a medical man. By far the most important sign of death is the cessation of the circulation and of respiration; when this is conclusively established, death is certain. Amongst other signs, but of distinctly less value, may be mentioned insensibility and loss of body heat, and also certain minor signs in reference

THE EARLY SIGNS OF DEATH.

to the state of the skin, the condition of the eye and the countenance.

1. *Cessation of Circulation and Respiration.*—It is desirable to group these together, because it is obvious that, if a person were still breathing, he could not be accounted dead because the action of his heart could not be detected; still less could a person whose heart could still be felt to pulsate be said to be dead on the strength of the apparent cessation of respiration. The two signs must therefore both be established before a person is pronounced to be dead. As a matter of practical experience, however, it will be found that the cessation of the heart's action affords more reliable evidence of death than the absence of breathing, because the means of detecting any pulsation of the heart are much more delicate than those methods by which movements of the respiratory apparatus are appreciated. The absence of the pulse at the wrists cannot be accepted as conclusive or sufficient evidence that the heart is not acting. If after careful auscultation for five minutes continuously over the region of the heart in a room quite free from noise the sounds of the heart cannot be detected, and if at the same time no signs of breathing can be perceived, it may be stated without hesitation that the person is dead. The tests to be adopted for determining the presence or absence of breathing are: auscultation of the lungs—the result of this method does not, however, afford such certain evidence as in the case of the heart; the holding a mirror in front of the mouth and nose—if any trace of moisture be present it will condense on the mirror, making a dull mark. This is very transient, the person therefore who performs this test must never take his eyes off the mirror. It is of more value as positive evidence of life than as negative evidence of death. Instead of

Cessation of circulation and respiration.

a mirror a feather may be held up in front of the mouth and nose; the same precautions as to watchfulness would be necessary, and the value of the evidence would be of the same kind. The last test that need be noted is the placing a glass of water or mercury on the chest; the slightest movement of the chest-wall would be betrayed by a ripple on the surface of the water or mercury. Amongst the other signs of the cessation of the heart's action, the most important is emptiness of the arteries; this can be verified, as M. Bouchut has shown, by an examination of the fundus of the eye with the ophthalmoscope. Besides the fact that the arteries are empty, which is shown by the extreme difficulty of recognising their position at all, it will be noticed that the veins are thready, and that they present a beaded appearance, due, according to M. Bouchut, to the disengagement of the gases in the blood at the moment of death. Again, the emptiness of the arteries may be ascertained by direct examination, when the vessels will be found pale, yellowish, and empty. The application of a ligature to the finger during life is followed, as every school-boy can testify, by the end of the finger so constricted becoming at first red, and ultimately bluish-red, and after removal of the ligature the constricted part is white. In the dead body these changes do not take place. After death no blood can be obtained by wet cupping, unless the test be applied immediately. A Frenchman has suggested that in a doubtful case an incision should be made through one of the intercostal spaces near the heart, and the finger inserted to feel if the heart's pulsations can be recognised!

Insensibility.

2. *Insensibility and Loss of Temperature.*—The mere fact of insensibility taken by itself would of course not be worth much as a sign of death. In many forms of brain disease, and in fainting, the

THE EARLY SIGNS OF DEATH.

insensibility may be complete, but a careful examination should reveal the continuance of the circulation if that were really going on. The most important case that has been recorded in relation to this subject is that of a gentleman who informed his medical attendant that he possessed the power of dying at will, and invited him to come with others to witness the experiment. So far as the doctors could discern, the experiment succeeded admirably, for during the space of more than half an hour they were unable to detect any sign of life, and were about to give the patient up for dead, when he began to breathe again. This case, however, occurred many years ago, and it is certain that, had auscultation then been known, the heart's action would have been discovered to be only enfeebled, not stopped. Under this head, too, might be mentioned cases of trance, catalepsy and prolonged sleep. The latter is not uncommon in certain forms of brain disease, the two former are occurrences infinitely rare in this country, but not infrequent amongst some continental nations. A few years ago, at Naples, a woman was reported to have been buried in a trance, and the doctor who signed the death certificate, and the mayor who ordered the interment, were sentenced to three months' imprisonment for 'involuntary manslaughter.'

It does not require any profound knowledge of physiology to understand that after death the body will tend to become cool. As to the length of time required for the process, several conditions will exercise an influence: the cause of death, the time of death, and the conditions under which the body has been since death will all play a part. It has been found that, assuming the temperature of the body at the time of death to be 85°, the rate of cooling will be one degree per hour, and the body will occupy from

Cooling of the body.

Rate of cooling. fifteen to twenty hours in cooling down, if exposed. A body will cool more rapidly in the open air than in a dwelling, and more rapidly in water than in air. The body of an adult cools more slowly than that of a child or an old person; the body of a lean person more rapidly than of one who is corpulent. In cases of sudden death and acute disease the cooling is retarded. The rate at which the body cools down is sometimes of great importance in cases of murder, as affording probable evidence as to the time of death. Thus, in the case of Hopley, the schoolmaster who was convicted of flogging one of his pupils to death, there was good reason to believe that the boy had actually died during the beating. The accused stated before the coroner that he went into the deceased's bedroom about six o'clock in the morning and found him dead, his body cold and his arms stiffening. It was proved that the prisoner had been beating the deceased up to 11.30 P.M. on the previous night, and as the body was cold when found, and rigidity was commencing, there was a strong probability that the deceased must have been dead at least six or seven hours, and therefore at a time when the prisoner was last known to have been with him. The body was well developed, covered with bedclothes, and the temperature at the time not low. The conditions all agreed well with the idea that the boy had died shortly before midnight. In another way the cooling may afford valuable evidence, and that is in the case of two persons found in the same place dead from violence. The presumption will be that that body which is the most cold has been the longest dead. The question of murder and suicide will have, however, to be decided more by the nature of the wounds, as it by no means follows that the person who died first was the murdered party. The murderer might have inflicted a more rapidly fatal wound on himself than on his victim. In one case where a man and woman were both found dead, the body of the woman was cold, and her wounds were such as could be more easily accounted for on the hypothesis that they had been inflicted by someone else, whilst the body of the man was still warm, and he had died from a wound in the throat which must have proved rapidly fatal, and which therefore in all probability was inflicted after the death of the woman.

The rule that a body commences to cool down immediately after death does not universally hold good. The temperature sometimes maintains a very high level, or even shows a tendency to rise. This has been observed after death from yellow fever, cholera, and even in cases of rheumatic fever. Putrefaction cannot be called into account for the rise in these cases, as sufficient time has not elapsed for its commencement. A person is not necessarily dead because the body is somewhat cold. The surface temperature may sometimes be found to be several degrees below that of the interior, as shown by a comparison of the temperature of the skin with that in the rectum. On the other hand, under some circumstances the body may sometimes retain a considerable amount of heat for an unusually long period. Thus, in a case recorded several years ago, which has been quoted very often, two practitioners went by order of the coroner to examine the body of a servant-girl found dead in her bed. Though the girl must have been dead for some eight or ten hours, the temperature of the body, it is stated, was not in the least diminished (thermometric observations not given). There was no sign of the commencement of rigor mortis, and all the organs were perfectly natural, and this fact, coupled with the warmth of the body, caused the practitioners in question to have some fear that the girl had not really been dead. The explanation of such a case as this would probably be found to be that the body had been covered by bed-clothes and lying in a warm room.

Pallor.

3. *The State of the Skin, Eye, Countenance, &c.*—The pallor of death is proverbial; nevertheless, it cannot be relied upon as a sign of death. There are many conditions during life under which extreme pallor of the skin is noticed, and sometimes after death the skin retains its natural colour for a while. It has been observed that, if the hand be examined after death by transmitted light, it appears marble-like and opaque, whilst during life it is transparent and rosy. The effects of heat or any caustic on living and dead skin may here be noticed, not that they would have much value as tests to determine whether a person was really dead or not, but because, in the case of a dead body found with marks of burning upon it, it might be of the utmost importance to decide whether those burns had been inflicted before or after death. When the flame of a candle is applied to a portion of skin, a blister is formed; if the person was dead at the time, or if that particular portion of skin was dead, the blister will contain a thin, non-albuminous fluid, and when it is snipped the skin below will have a dry and glazed appearance. On the other hand, a blister formed during life will be found to contain a highly albuminous fluid; the subjacent skin will be reddened, and in a day or so there will be formed a red line of demarcation around the blister which is never seen in the case of dead skin, and is an absolute proof of the vitality of the part. The application of a caustic to a living person produces a brown or black discolouration with a zone of redness, whilst in the dead body its application has little or no effect.

Effects of burning on living or dead skin.

Cadaveric ecchymoses.

Cadaveric lividities or *ecchymoses* are the purple

or reddish-purple stains that appear on the surface of the body after death. Their importance is very great, and for this reason: that by a casual observer they <u>might be mistaken for bruises</u>, and therefore might give rise to a totally groundless suspicion of violence. They form during the cooling of the body, and usually precede the onset of rigidity; they may be found anywhere over the body, but not on parts that have been subjected to pressure. They are generally <u>best seen in those who have died rather suddenly</u> from a comparatively vigorous state of health; they may be of any shape and size, their margins are usually abrupt, and they do not shade gradually off into the surrounding skin. <u>They are due to stagnation of blood in the minute capillaries, and not to extravasations, as might have been supposed.</u> This is proved by the fact that, on cutting through one of these patches, the deeper layers of the skin are not found to be infiltrated with blood as they would be in an ordinary ecchymosis. True bruises may be distinguished from these by their seat, the cadaveric lividities occurring mostly in the dependent parts; also by their shape—bruises due to an extravasation of blood will be swollen, and on section blood will escape from them. Moreover, a bruise passes through various shades of colour, and is always surrounded by a zone of colours shading it off gradually to the healthy skin. Corresponding changes may take place in the internal organs at this period,' and patches of ecchymosis are found on the posterior aspect of the brain and spinal cord, the posterior part of the lungs, and the dependent parts of the stomach and intestines.

As regards <u>*the eye*</u>, after death the pupil becomes moderately dilated, and of course no longer responds to the stimulus of light. For some time after death, certainly half an hour and probably longer, how- {The state of the eye.}

ever, the pupil dilates or contracts under the influence of atropine or eserine respectively. The iris, like all the other muscles of the body, becomes flaccid after death, and if pressure be applied to the eyeball the pupil can be made to assume an oval or any other shape, and to retain it. During life it is exceedingly rare for the pupil not to regain its normal rotundity immediately the pressure is withdrawn. As a rule, the cornea speedily becomes hazy after death, and, owing to the absorption of the aqueous humour, it also becomes wrinkled. In cases of poisoning by the oxides of carbon and compounds of cyanogen, and in some cases of apoplexy, however, the cornea retains its transparency for a long time, and in cholera and certain other diseases it may become cloudy during life. The other changes in the eye are some of them so obvious as hardly to need mention—such, for instance, as anæsthesia of the conjunctiva. The conjunctivæ soon become turbid after death, as also do the various media of the eye. As a general rule, after death the eye sinks in the socket, and, owing to the absorption of some of the media that has already been referred to, it becomes soft, so that when pressed with the finger it retains the indent. When putrefaction has set in, the development of gases in the orbit may cause the eye to protrude. The countenance usually retains the expression which it had at the time of death, and therefore is mostly described as being placid; but in death by violence the features often bear witness to the mental state of the deceased during the final struggle.

The muscular system. The state of the *muscular system* alone remains for consideration. During this pre-putrefactive period the muscles pass through two stages, the first of relaxation, the second of rigidity; the former lasting, under ordinary conditions, three hours or thereabouts, the latter from twenty-four to forty-eight hours.

THE EARLY SIGNS OF DEATH.

The duration of the period of *flaccidity*, as just stated, is about three hours, but it is liable to considerable variation, being much influenced by the cause of death, the age of the person, and the season of the year. All the muscles of the body during this stage are soft and flabby, so that the lower jaw drops, the pupil becomes dilated, and the sphincters are no longer tightly contracted. Whilst this stage lasts the muscles can be excited to contraction by the interrupted current, and even by chemical and mechanical stimuli.

As regards the order in which this muscular irritability disappears, Onimus states that the first muscles to lose it are the diaphragm and the tongue, then come the facial muscles, the masseters being the last of these. In the limbs the extensors go before the flexors, five or six hours after death the muscles of the trunk still show some excitability, and the abdominal muscles may respond to faradism even later than this. Nysten, another authority who investigated this point, states that the irritability ceases first in the left ventricle of the heart, then in the intestines, stomach, bladder, right ventricle of the heart, œsophagus and iris, afterwards in the muscles of the trunk, and of the upper and lower extremities, and lastly in the right and left auricles of the heart. From some experiments made in one of the Paris hospitals on the bodies of patients recently dead, it was ascertained that muscular irritability ceased in the trunk and limbs in death from peritonitis in three hours, from phthisis and cancer in from three to six hours, from heart disease and hæmorrhage after nine hours, from paralysis in twelve hours, from pneumonia in from ten to fifteen hours. The contractility is influenced too, it should be borne in mind, by exposure of the body after death to the influence of certain gases or vapours. Thus ammonia, sulphuretted hydrogen, and carbonic

Extinction of muscular irritability.

anhydride diminish the duration and intensity of the contractility, while carburetted hydrogen, chlorine and sulphurous acid are said to be without any influence. This stage has never been known to last so long as twenty-four hours, but it sometimes is only of a few minutes' duration; it ceases when rigor mortis sets in.

Rigor mortis. The period of muscular rigidity, or *rigor mortis*, sets in immediately on the termination of the last stage, and persists for one or two days. It passes off with the onset of putrefaction. The time at which it comes on and the duration of its existence are neither of them fixed periods. Thus it is more complete in the bodies of old people, and in infants it usually sets in very early. A low temperature favours rigidity, and therefore in winter it generally persists for a long time. It is better marked in the bodies of muscular people than in those who are emaciated. When death has supervened on great exertion or fatigue, rigor mortis usually passes off soon. In death from lightning it sets in and passes off very speedily; so quickly, in fact, that the popular notion is that there is no stage of rigidity at all. Rigor mortis takes place both in the voluntary and involuntary muscles, and it commences in those muscles that retain their heat the longest. The exact cause of it has been the source of a good deal of dispute.

From the facts, that in hemiplegia the paralysed limbs are not differently affected by it from the non-paralysed, and that neither division of the nerves nor removal of the brain influences its onset, it is evident that the nervous system plays no part in its production. Some have thought that the coagulation of the blood had something to do with the causation of rigor mortis, but as it may be found when the blood still remains fluid, it is clear that this cannot be the true explanation. It is probable, therefore, that rigor mortis is due to a local condition of the muscles. The stiffening of the muscle which takes place during rigor mortis is apparently dependent upon the formation of myosin as a result of the coagulation of muscle-plasma. The muscle becomes opaque, and whereas its reaction in the normal state is alkaline, when in this state it is acid.

Formation of myosin.

During rigor mortis the muscles retain the exact position they occupied at its commencement. Rigor mortis usually first shows itself in the eyelids and lower jaw, next attacking the other muscles of the face, then the legs, and lastly the arms. It passes off in the same order as it comes on. It is usually complete within two or three hours of its onset. Most commonly it appears about three hours after death, but it may come on very much earlier. Thus Brown-Séquard observed a case in which it set in three minutes after a man had stopped breathing, and before the heart had actually ceased to beat. In less than an hour putrefaction had commenced. Rigor mortis commences in the involuntary muscles sooner than in the voluntary, and the heart is often affected in an hour. Sir James Paget has pointed out the danger of mistaking this contraction of the heart for a state of hypertrophy, and the subsequent flabbiness for a condition of dilatation. Rigor mortis is late in appearing in the bodies of healthy people who have died suddenly, and had not undergone any great fatigue just before death.

There is one condition which requires especial notice here—namely, what has been called *cadaveric spasm*. In some cases of violent death, and under certain other conditions, the body appears to pass into a state of rigidity at once, no stage of flaccidity being perceptible, and thus retains the position in which it lay at the moment of death. This is seen in the bodies of soldiers killed at the end of a long day's fighting, whilst those who died early in the battle, in conformity with what has been stated

Cadaveric spasm.

above, become rigid very slowly. Authors are not
agreed as to the nature of this spasm. Taylor, for
instance, does not regard it as true rigor mortis,
whilst Tidy contends that it does not differ from
rigor mortis in any essential particular. As it is
said to pass gradually into rigor mortis, there does
not seem to be any necessity for regarding it as
other than a form of rigor mortis. Early rigidity
is often of great value, as affording evidence of
suicide or murder, as the case may be. For instance, if a weapon is found so grasped in the hand
of a dead man that it can only be removed by the
exercise of some force, and if there are present
wounds or a wound such as might have been inflicted by that weapon, then the presumption will
almost amount to a certainty that the wounds were
self-inflicted. In one case a gentleman was found
dead in his bedroom shot through the head; a discharged pistol was firmly grasped in his right hand,
and there was a loaded one in his left. There could be
no question as to this being a case of suicide. Sometimes murderers have, after the crime, placed the
weapon with which it was committed in the hand
of their victim, hoping thereby to raise the idea of
suicide. Very often in these cases the weapon has
been put into the left hand of a right-handed person
or *vice versâ*; or, in the case of a knife, the back of
the knife has been turned the wrong way, so that
the imposture is recognised at a glance. But,
apart from such direct evidence, the deception is
certain to fail, for the simple reason that the weapon
will be lying loosely in the hand of the deceased,
a fact in itself sufficient to negative the idea of
suicide. The great importance which may attach
to the position of the weapon will be evident from
the following case quoted by Taylor:—A medical
man at Bordeaux was called upon to examine the
body of a gentleman supposed to have been mur-

Margin: Its value in cases of violent death.

dered. He found the deceased perfectly dead, sitting in an arm-chair by the side of the bed, the left elbow resting upon a bolster. The right hand, which held a recently-discharged pistol, rested upon the middle of the right thigh, the greater part of the barrel projecting over the inner surface of the thigh, so that the slightest motion of the part would apparently have been sufficient to cause it to fall on the floor. The temperature of the body indicated that the deceased had not been dead above two hours, and it was at about that time that the neighbours had heard the report of a pistol. The bullet had traversed and fractured the left parietal bone, and the deceased had lost a large quantity of blood. The other facts ascertained relative to the case were that the deceased, who was about 60 years of age, had never shown any disposition to destroy himself, and there was no circumstance which seemed likely to have acted as an exciting cause, except perhaps the loss of a lawsuit by a favourite sister, which, however, seemed scarcely a sufficient explanation. He had a son, who lived in the house with him and slept in the same room. They were both dissipated in their habits. On the morning in question the son, according to his own statement, after breakfast threw himself on his bed, which was beside that of his father, and fell asleep till he was aroused by the discharge of a pistol. He was accused of having murdered his father, and placed the pistol in his hand to simulate suicide. The police authorities took this view strongly, not only because there were reasons why the son might have wished to get rid of the father, but also because they found that on raising the hand of the deceased to the position in which it must have been when the shot was fired, and then allowing it to drop, on every occasion the pistol fell to the ground. The medical man at once pointed out that these experiments were of no value, and that the retention of the weapon in the hand was an evidence of its having been held in the hand during life and at the moment of death, and that it afforded the best possible proof that the case was one of suicide. Had it not been for this evidence the son would in all probability have been tried for the murder of his father.

Occasionally, much more often in other countries than in this, public feeling has been strongly aroused by the statement that a person has been buried alive whilst in a trance or in some such state as would closely simulate death. M. Bouchut has devoted a considerable portion of his book ('Traité des Signes de la Mort') to explaining the means taken to prevent premature burial, and to the arguments for and against those measures. Some years ago a Frenchman stated that he had known of 46 cases of premature interment. Of these, 21 persons returned to life just as they were about to be deposited in the ground; of the remainder, nine recovered owing to the affectionate attentions of their relatives, four from the accidental falling of the coffin, two from a feeling of suffocation in their coffins, three from the punctures of the pins in fastening the shrouds, and seven from unusual delay in the funerals! He then drew up a table of the yearly deaths in Paris, and from a comparison with these cases of his pointed out how many thousand people a year were buried alive in Paris alone. It is difficult to admit even the bare possibility of such an occurrence as premature burial. If ordinary care be taken, such a thing would be impossible. No one should be pronounced dead until it is certain that the heart's action has stopped for some little time, and if there was the least doubt in the mind of the medical man as to whether the heart had really stopped or not, he ought to make a fresh examination in an hour's time.

Alleged premature interment.

CHAPTER II.

THE LATER SIGNS OF DEATH.

THE more remote signs of death are those due to chemical changes announcing the dissolution of the organic compounds of the body. Putrefaction is defined to be 'a spontaneous change common to all nitrogenised organic bodies when exposed to the air, whereby they become resolved into new and simpler products' (Tidy). It generally commences when rigor mortis passes off, or, to speak more correctly, rigor mortis passes off when it supervenes. In this country, therefore, in ordinary circumstances it generally commences about the third day after death, but no hard-and-fast rules can be laid down on this point. To be an evidence of death the putrefaction must be general, as partial gangrene may, as is well known, occur during life. As an instance of most unusual retardation of putrefaction, the following case, given by Taylor, may be quoted. A young man who had recently recovered from an attack of ague, died with symptoms of phthisis, but not of a well-marked character. The cause of death appears to have been obscure, and in order to satisfy any doubt as to the reality of death, some wounds and cauterisations were made on the body, and on the third and fourth days these had, it is said, passed into a state of suppuration. On the fifth day his right hand turned black

Putrefaction.

The date of its appearance.

and closed; from the fifth to the ninth days a clammy perspiration was perceived upon the skin, and some vesicles containing serum were formed on the skin of the back. During this time there was no appearance of respiration or circulation, and the limbs, although cold, were pliant and flexible. The forehead was furrowed with vertical wrinkles, and the countenance had an expression not usually observed in a dead body. On the 18th day the lips presented their usual red colour, and although the body was lying during this time in a warm room, there was no cadaveric odour or ecchymosis. Signs of putrefaction were seen on the 20th day. In another instance a body retained its natural appearance for 28 days, and even on the 35th day putrefaction was but little advanced.

Putrefactive lividity. One form of lividity seen after death, and due to the stagnation of blood in the minute capillaries of the skin, has already been mentioned, but there is another form due to incipient putrefaction. This is seen chiefly on the dependent parts of the body, which tend to become of a deep red colour on this account. On cutting through a patch of skin thus discoloured, it will be found that there is an infiltration of blood throughout its layers. This condition has been named suggillation by some authors; the parts affected gradually undergo changes of colour, ultimately becoming green. Putrefaction usually first shows itself in the skin of the abdomen, especially just round the umbilicus, by a greenish discolouration; this gradually spreads to the scrotum, thorax, neck, face and extremities, these parts becoming brownish in colour and gradually getting darker. Very soon after the commencement of these changes in the skin, gases begin to be developed in the internal cavities, and later on in the subcutaneous tissues, leading to great distension of the abdomen, scrotum, face, and swelling about the eyes, so that the person is quite unrecognisable. One result of this development of gas is particularly worth mentioning, and that is, that if there exist any surface wound the pressure of the gas is liable to drive any blood in its neighbourhood through it. This post-mortem bleeding, as it has been called, was regarded with much awe in bygone days, and it was commonly believed (and the belief was acted upon) that if a person had been murdered, his wounds would bleed when his body was touched by the murderer. Many a criminal must have escaped his just deserts when this test was regarded as affording conclusive evidence of his guilt or innocence. Besides the emptying the heart and large vessels of blood, the presence of frothy mucus and even of food about the mouth and nostrils may be mentioned. Cases have been recorded, too, in which a fœtus has been expelled from the gravid uterus by the distension of the abdomen with gas, and in one instance it is said that a living child was thus born. The gases to which these results are due are sulphuretted hydrogen, ammonia, nitrogen, carbonic acid, and carburetted and phosphuretted hydrogen. As regards the internal viscera, the larynx and trachea are the first to show the effects of putrefaction. The changes in the mucous membrane as a consequence of putrefaction are the same throughout the body, but their presence in the stomach is of the greatest importance from a medico-legal point of view. Redness of the lining membrane is the earliest sign of decomposition in the stomach, and the fact that it shows itself in the most dependent parts affords the means of distinguishing with certainty this state from that due to poisoning. Redness is also to be seen in the dependent parts of the intestines. On section it will be found that the redness extends through all the coats of the stomach or intestine, and can

Post-mortem bleeding.

Putrefaction in internal organs.

be recognised on the peritonæal aspect as a rule. Redness due to inflammation, passive congestion or poisoning does not generally affect all the coats, or if it does, it affects the lining membrane much more severely than the others.

<small>Post-mortem digestion.</small>

Post-mortem digestion of the stomach or cadaveric softening must not be altogether omitted. It is due to the softening and ultimate giving way of the coats at one point, sometimes leading to perforation, and the passage of the contents into the peritonæal cavity. It is most commonly found at the posterior part of the greater curvature, and the appearances will depend upon the extent to which the various coats have been involved. In a slight degree the rugæ will have disappeared, and the surface will appear to be smooth; when the softening has extended deeper, large vessels will be seen in the submucous tissue, their course being marked out by ill-defined black lines. When perforation has taken place the edges of the rent, which is large, will be found thin, dissolved, flocculent, ragged and hanging in shreds. (For further description of cadaveric softening, the reader is referred to Wilks and Moxon's work on Pathology, from which the above account has been taken.) This condition may be distinguished from perforation due to ulceration during life, by the absence of any thickening about the edges, as in the case of a chronic ulcer, or of any inflammatory signs, as in an acute case. Further, had the perforation taken place during life, there would certainly have been set up some degree of peritonitis.

The parts in the immediate neighbourhood of the gall-bladder become stained with bile. In the heart, putrefaction shows itself by a deep redness of the endocardium, especially that lining the valves. The aorta and pulmonary artery, too, show a deep uniform redness of their lining membrane.

Softening is the main feature of decomposition in the internal viscera. There should be no difficulty in attributing it to its right cause; inflammatory softening, when taking place during life, does not often affect the whole of an organ, and even if it does, as happens in the case of the spleen in typhoid fever, the existence of softening in other organs at the same time should serve to prevent any mistake. The brain in children undergoes putrefaction very early, sooner than in adults; the lungs are late in being affected, but the uterus resists putrefaction longer than any of the other viscera.

It is often very desirable to form an approximate idea as to how long life has been extinct in the case of a dead body in which putrefaction has commenced. There are many considerations which have to be borne in mind in endeavouring to come to a conclusion on this point, and the more important of them will be referred to directly, but the following rules suggested by Casper, and somewhat modified by Tidy, may be helpful:—'*From one to three days after death* : A light green colour, visible about the centre of the abdomen ; the eyeballs soft, yielding to external pressure. *From three to five days* : The green colour of the abdomen becomes intensified and general, spreading if the body be exposed to the air or buried in the ground, in the following order—genitals, breast, face, neck, superior and inferior extremities. *From eight to ten days*: The colour becomes more intense, the face and neck presenting a shade of reddish green. The ramifications of the subcutaneous veins on the neck, breasts, and limbs become very apparent; the patches congregate ; gases begin to be developed in and to distend the abdomen and hollow organs, and to form under the skin in the submucous and intermuscular tissues; the cornea falls in and becomes concave; the sphincter ani relaxes; the

Inference of date of death.

nails remain firm. *From fourteen to twenty-one days*: The colour over the whole body becomes intensely green with brownish-red or brownish-black patches; the body generally is bloated, from the development of gases in the abdomen, thorax and scrotum, and also in the cellular tissue of the body generally; the swollen condition of the eyelids, lips, nose and cheeks is usually of such extent as to obliterate the features, and destroy the identity of the body; the epidermis peels off in patches, while in certain parts (more particularly over the feet) it will be raised in blisters filled with a red or greenish fluid, the cuticle underneath frequently appearing blanched; the colour of the iris is lost; the nails easily separate; the hair is loose. *From the fourth to the sixth month*: The thorax and abdomen burst, and the sutures of the skull give way from the development of gases within; the viscera appear pulpy, or perhaps disappear (melt away), leaving the bones exposed (colliquative putrefaction); the bones of the extremities separate at the joints.'

<small>Modifying conditions:—External.</small>

The conditions which influence the development and rate of putrefaction may be divided into those dependent on the body itself and those which are independent of the body. The latter are by far the more important. They are three in number. 1. <small>Temperature.</small> *Temperature.* Putrefaction proceeds most favourably when the temperature is between 70° and 100°, and not at all when the body is frozen (32°), or exposed to a very high temperature (212°). In either case putrefaction is arrested, but in the case of the frozen body, if it is afterwards exposed to the ordinary temperature, decomposition proceeds with unusual rapidity. The presence of light does not make <small>Moisture.</small> any difference. 2. *Moisture.* This is the most important of the three; without it putrefaction cannot take place at all. In an excess of moisture, however, it does not probably occur, owing to the fact that the access of air is thereby prevented. Substances which have no moisture in themselves do not undergo putrefaction; witness the hair, teeth, bones and nails. 3. *Air*. This is almost as essential as moisture; bodies protected from the air decompose very slowly. It is the oxygen in the air that plays the chief part, neither nitrogen nor carbonic acid assist in the process at all. Stagnant air is more favourable for decomposition than air in motion. *Air.*

The conditions within the body which influence the development or duration of putrefaction are several. The bodies of children decompose more rapidly than those of old people, and old people more rapidly, according to some authorities, than adults. The bodies of females are said to decompose more rapidly than those of males. It is said that peculiarity of constitution has some influence, but how this is possible has not been shown. From what has already been said about the effects of moisture, it is obvious that a lean body will be less prone to putrefaction than a stout one. Decomposition sets in early in a wounded part or a part inflamed during life. As a general rule, acute diseases are accelerators of putrefaction, and chronic diseases retard it. Thus in pyæmia, after puerperal fever, typhoid, scarlet fever or diphtheria, putrefaction sets in early and proceeds rapidly. In chronic diseases too, associated with dropsy, putrefaction sets in soon: here again the influence of the moisture is probably the determining cause. The bodies of plethoric but healthy people who have died suddenly are said to putrefy slowly. Some poisons accelerate decomposition: such are prussic acid, morphia and the narcotics generally, and strychnia when death has been much delayed. The explanation of these facts appears to be that when death has really been *Modifying conditions:— Internal.*

due to exhaustion, no matter whether from repeated convulsions or any other cause, rigor mortis and putrefaction are hastened. Arsenic and chloride of zinc retard putrefaction, and so practically do lime and the mineral acids, although the popular belief on the subject is to an exactly contrary effect. Acting on this fallacy, murderers and others have occasionally used lime and other chemicals in the hope of completely destroying the most important evidence against them in the shape of the dead body, but have only succeeded in so preserving it as to render the cause of death ascertainable beyond all dispute. Such a result took place in the body of Ann Palmer, murdered by Palmer, and in the victims of Wainwright and the Mannings.

Putrefaction in the earth. The changes which take place in the body in the earth must be considered next. The early changes are the same, the internal viscera become soft and shrink. After a time there appear on the surface of some of them, especially the liver, small white patches of incrustation of a crystalline substance soluble in water, which consists of phosphate of lime, mixed with phosphate of ammonia and magnesia. The body gradually becomes covered with a fatty deposit, and the soft parts undergo gradual saponification, and finally disappear, leaving only the long bones, the bones of the base of the skull, the short flat bones and the vertebræ having disappeared. As already mentioned, the hair and the teeth remain. Those circumstances which have already been mentioned as influencing, in one way or the other, putrefaction in the air, will exercise a similar influence in regard to bodies in the earth; but, in addition, other points must be noted as *Modifying conditions.* capable of having a great effect. For instance, it makes a great deal of difference how soon after death the body is buried; the longer it is kept above ground the more rapidly will decomposition pro-

THE LATER SIGNS OF DEATH. 31

ceed. When putrefaction has set in, there are no less than twenty species of larvæ that deposit their ova in the body; these, of course, materially hasten decomposition. Again, much will depend upon whether the body is put into the ground as it is, or in a coffin, and if the latter, the nature of the coffin is important. A body put naked into the ground will of course decompose much more rapidly than one enclosed in wood, and the latter more rapidly than one in a leaden coffin. Bodies buried in valleys decompose more quickly than those buried on high ground; a soil of clay or marl favours decomposition, sand, gravel or chalk retard it. The fluctuations of temperature between the day and night affect the ground to a depth of three feet; those due to the seasons of the year make themselves felt to a depth of six feet; below this level the temperature of the earth is tolerably uniform. It follows, therefore, that bodies buried less than six feet below the surface will be influenced by the changes in temperature, and will therefore be liable to more speedy decomposition than those which have been interred below this level.

As regards the estimation of the time that may have elapsed since death in the case of a body buried in the earth, Tidy makes the following statements, based on the observations of Orfila: After periods varying from a *few months* to *one and a half to two years*, it will usually be found that the soft tissues of a body buried in a coffin become dry and brown, and the limbs and face covered with a soft white fungus. About this time hard white crystalline deposits of phosphate of lime form on the surface of the soft organs. If these crystals occur on the mucous membrane of the stomach, they might be mistaken for the effects of poison. After a period of *four years*, the viscera become so mixed together that it is difficult,

Inference the date of death.

if not impossible, to recognise or to distinguish them. After *eight years'* burial, Dr. Taylor found a body in fragments, the soft parts being loosely adherent to the bones and covered with a white, fibrous, offensive substance. The features were entirely destroyed and the bones of a dark colour. At later periods the soft parts, as a rule, entirely disappear. If a body has been buried in a very dry soil, portions may be found brown and mummified, in which condition they remain unchanged for a long time. In course of time the bones become disarticulated. At a still later period the long bones only remain, the short flat bones, the base of the skull and the vertebræ crumbling to a white powder. The bodies of children decay more rapidly than those of adults. If a body be placed in a thin wooden coffin in a moderately superficial grave, it is probable that at the end of *ten years* the soft parts will have completely disappeared.

Adipocere.

After death, one of the changes that may go on in the tissues is that known as *saponification*, which results in the formation of adipocere. This last is a whitish or yellowish brown soapy substance, having a rancid odour when warmed, and being hard, white and brittle when dry. It was first described about a hundred years ago by a Frenchman named Fourcroy, who often found, on opening graves, nothing but irregular masses of soft ductile matter of a grey-white colour, sometimes brittle and dry, but always unctuous and soapy. It is a true ammoniacal soap—that is, a combination of certain fatty acids with ammonia for a base. As ammonia is an essential constituent of adipocere, this cannot be formed from pure fat alone, which contains no nitrogen; but as fibrin contains nitrogen, and as fat cells are enclosed in a membrane which is composed of fibrin, it follows that, practically, wherever fat is, there the formation of adipocere can take place. And that this is so is amply proved by the fact that the formation of adipocere goes on more readily in fat bodies than in lean ones, and in children than in adults, the amount of superficial fat in the former being always so much in excess of that present in the latter. Other circumstances which favour the formation of adipocere are burial in a deep grave, or in an overcrowded churchyard, or in a cesspool; the presence of running water also encourages the formation of it. As regards its chemical characters, it floats on water, burns with a bright yellow flame, and when heated on platinum it gives out a smell of ammonia; the ash that remains after burning consists of soda, potash, phosphate of lime, and oxide of iron. When mixed with water it forms an acid, opaque, leathery compound. The breasts, cheeks, and kidneys are soonest converted into adipocere, the muscles resisting longer. Devergie gives three years as the date for its formation when the body is covered with moist earth, and one year when it is in water. According to Casper, adipocere is not formed to any great extent under three or four months' submersion in water, and six months' burial in moist earth. *[margin: Circumstance favouring its formation.]*

Putrefaction takes place more slowly in the water than in the air or in the earth. This is partly on account of the deprivation of air, and partly because the water is colder. Running water retards putrefaction more than still water, and a deep pool more than a shallow one. The first effect of the water is to produce a sodden condition of the skin of the hands and feet. A body, when taken out of the water after a few days' immersion, may show scarcely any signs of decomposition, but within a very few hours of its removal decomposition will set in, and proceed with great rapidity. It is noteworthy that, owing to the saturation of *[margin: Putrefaction in water.]*

the skin with water, bruises are, at the time of the removal of the body from the water, quite imperceptible, but in a short time, when some of the water has been got rid of, they begin to make their appearance. The changes in colour from putrefaction, in the case of a body in the water, commence in the face and neck, and spread thence to the chest, trunk, and limbs. When the body has been a few days in the water it rises to the surface, owing to the development of gas in the abdomen; the gas escapes after a varying period, and the body sinks, to rise again after a fresh accumulation of gas. The time after death at which this rising may be expected to take place will be considered in the chapter on Death from Drowning. When it rises to the surface the highest part is either the abdomen or back, according as the body happens to be on its face or not; it is said that in women the abdomen is always uppermost. Various reagents have been suggested to restore the features of a drowned person; chlorine and the chlorides figure largely amongst them. Tidy speaks strongly in favour of a saturated solution of alum and nitre in alcohol.

Inference of date of death.

According to Devergie, the following may be relied upon in determining the period of death in the drowned. *First four days*: Rigidity may persist, especially if the water is cold. *Fourth or fifth day*: Skin of fingers begins to whiten; this spreads to palms, and later to soles. *Fifteenth day*: A greenish-red spot begins to form about centre of sternum; skin of palms is wrinkled; towards the end of the month the cortical substance of the brain is greenish. *One month*: Face reddish brown, eyelids and neck somewhat green, brown spot on centre of sternum about six inches in diameter, skin very wrinkled, but hair and nails still adherent, the scrotum and penis much distended.

Sixth or seventh week: Neck and thorax very green; cuticle begins to separate at wrists. *Two months*: Body covered with green slime; face much swollen; skin comes off from hands and feet, with the nails, like a glove; if at the time of death the right side of the heart had been engorged, the lining membrane of it will now be jet black. *Two and a half months*: Green colour of skin on arms, forearms, and legs; some adipocere on cheeks, chin, breast, and inner sides of thighs; muscles not much altered in colour. *Three and a half months*: Scalp, eyelids, and nose much destroyed; recognition impossible; skin of breast greenish brown; lungs do not fill the thorax, but leave a space lined with a reddish serum. *Four and a half months*: Skull bare; calcareous incrustation commencing in the adipocerous parts. These changes are the average for winter; in summer the changes would take place from three to six weeks earlier. Taylor has recorded a very important case in which a question arose as to the time required for the production of adipocere in the drowned. The case was as follows:—The property of a bankrupt, who had committed suicide by drowning, was seized under a commission, and an action was brought to recover it on the ground that the insolvent was dead when the commission was issued. The deceased left his home on November 3rd, and on December 12th following—*i.e.* five weeks and four days after his departure—his body was found floating in a river near the place where he resided. A commission of bankruptcy had been taken out against him a few days after he was missed, and it became therefore important to determine whether he had drowned himself (for there was no question as to his having committed suicide) before or after the date of the issuing of this commission. If it could be shown that he was already dead at that date, the commission would be void in law, and the property could not be seized under it. The litigation then turned upon whether he had drowned himself upon the day of leaving his house, or at some subsequent date. The body was found floating with the head and feet submerged, and the face was covered with a muddy slime. The body was discovered on a Wednesday, and a coroner's inquest was held on the following Saturday. On the day before the inquest three medical men examined the body, with a view to ascertain whether any changes had taken place in it, which would justify an opinion as to the length of time during which it had been lying in the water. The muscles of the buttocks were found to be converted into a fatty substance much resembling adipocere. The hair of the head separated from the scalp by a slight pull. The face was completely disfigured by putrefaction, the other parts of the body being firm and white and free from putrefactive change. The shirt and neckcloth were so rotten as to be torn by the slightest force. A medical witness for the plaintiffs stated it as his opinion that the body could not have been less than six weeks submerged. Three or four weeks would not have sufficed to produce the appearances met with. Another witness thought that the deceased's body must have been in the water during the whole time that he had been absent. The verdict of the jury was in accordance with the medical evidence, and was to the effect that the deceased was dead at the time when the commission was issued.

Mummification.
Mummification signifies the condition arrived at when the body has been protected from the air and moisture, as when buried in the sand or in a perfectly air-tight coffin. The body becomes dry and shrivelled up. It is not possible to say how long a body may remain in this state.

CHAPTER III.

IDENTITY.

IDENTITY in the case of the living does not of necessity call for scientific evidence, though such may be of great value in reference to any peculiarities about the person, such as scars, deformities, &c. But very often it has to depend upon certain peculiarities in manner, gait, or appearance, which the witness is able to swear to. Prison warders, and those who make it their business to study features, are generally able to speak with tolerable confidence as to the identity of a person whom they have seen before. In the case of a recently dead body the need for scientific evidence is greater; and when the features have become altered by decomposition, all the ordinary means available in the case of the living body are absent. The medical expert would then determine the sex, the age as near as he could, the probable occupation of the deceased, and examine the body for any marks, whether congenital or not, whereby the identity might be established.

Personal identity.

But it is in the case of the discovery of mutilated remains that scientific evidence is most essential. The first point to be considered, when in the presence of remains of a body, would be to determine whether they had all formed part of the same body. The absence of any duplicate parts ought not to be difficult to establish. The probable height of the

Mutilated remains.

deceased may be estimated with great accuracy by certain rules that will be mentioned presently. It is sufficient to say, with regard to their accuracy, that in the case of the murder of Dr. Parkman, though the head and neck and feet were missing, the experts who examined the remains determined the height of the deceased within half an inch of the exact measurement.

The cause of death. Injuries likely to have caused death, and the method of mutilation, are points that should not be overlooked: in more than one instance, by this means, a murder has been correctly traced to its perpetrator. To have ascertained the probable cause of death is to have gained knowledge that may be the means of bringing the crime home to some person or persons. In reference to this subject the following case is quoted by Taylor, and may be cited:—
In 1857 the remains of a human being were found in a bag on one of the buttresses of Waterloo Bridge; the bag contained some clothing as well as the remains. There were 23 portions of the body, the flesh having been roughly cut from the bones, and the latter cut and sawn into small lengths, probably with a view to reducing their bulk. The trunk was in eight pieces, the upper limbs in six, the lower in nine. The head, the seven upper dorsal and the cervical vertebræ were missing, as well as the hands, feet, and some portions of the left side of the chest. The internal viscera had been removed. The long bones were in their full state of development; they had been sawn through near the joints with a fine saw. On the left side of the chest, between the third and fourth ribs, there was a stab which had penetrated the thorax, and which, if inflicted on a living person, must have entered the heart; the edges of the wound were everted and wide apart, and the muscles around were infiltrated with blood. From the state of the soft

parts and joints, it appeared that the body had been cut up and exposed to a boiling temperature whilst the limbs were in a state of cadaveric rigidity. The conclusions arrived at were: That the body was that of a man 5 feet 9 inches high, probably of dark complexion; that the only evidence of violence was a stab in the heart region, which might have been inflicted before or after death; that the body had not been cut up by anyone with any knowledge of anatomy; and that the person might have been dead three or four weeks before the examination was made. The clothing found in the bag had belonged to a man, evidently a foreigner; the various articles were much torn, and in some cuts and stabs were found, and they were all more or less bloodstained. A stab found in the collar of the overcoat had evidently been inflicted with great force, as it had passed through the coat and waistcoat as well, and these had marks of blood on the inner side. There were also marks of blood on the inside of the clothes corresponding to the wound in the left side of the chest already alluded to. The examiners concluded that the state of the clothes was consistent with the idea that they were worn at the time of the murder. The identity of the deceased was never established, and no clue as to the murderer found. In this case, no doubt to destroy the identity, the remains after being cut up had been boiled. A similar plan was adopted by Kate Webster, who murdered an old lady named Thomas a few years ago at Richmond, but notwithstanding this precaution and the fact that the head was never found, the identity of the remains was fully established.

It is often essential, in order to bring the crime of murder home to a particular person, that the remains should be identified. This was well shown in the trial of Wainwright for the murder of Harriet Lane in 1875. The latter had been missing for

Case of Harriet Lane.

about a year, and when last seen had been in the company of Wainwright, by whom she had two children. A year after her disappearance, Wainwright was caught in the act of removing some human remains from one part of London to another. The remains were those of a female, and though the body had been dead apparently about a year, they had been recently cut up into small portions. Some parts of the body were adipocerous; others were in tolerably good preservation, owing to the action of chloride of lime, which had been put into the grave from which the body had just been removed. The marks of violence found on the body were amply sufficient to have caused death, and it was undoubted that the woman had met with a violent death. As the evidence was only circumstantial, it was of great importance, in order to connect the prisoner with the murder, that the remains should be identified as those of Harriet Lane, of whose disappearance he had already given conflicting accounts, and whose death would free him from a connection which, as a married man, he would be glad to be quit of. The body had been buried on premises rented by the prisoner, and in the grave certain articles were discovered which were recognised as having belonged to the missing woman. The features could not be recognised; the question of identity therefore depended solely on scientific evidence. The age of the deceased was 24; this coincided with the probable age of the body, in which three wisdom teeth were found. The height of the deceased was believed to have been 5 feet 4 inches; the remains measured 4 feet $11\frac{1}{8}$ inches. The difference between these, it was thought, might fairly be attributed to shrinking of the intervertebral discs. The colour of the hair was not quite the same as that of the missing woman, but some allowance was made for the probable action of the chloride of lime, and no stress was laid on this apparent discrepancy. One of the most important pieces of evidence was the discovery of a scar on the right leg just below the knee, exactly in the position in which a scar had been proved to have existed in the body of Harriet Lane. Even this, however, would have failed to establish the identity if it could have been shown that the uterus was that of a woman who had never borne children, an opinion which was maintained by one medical witness. The value of this opinion, however, was much reduced when the witness stated his belief that it was impossible to decide this point with certainty in any case. On the other hand, the witnesses for the Crown expressed a very positive opinion that the woman had borne children; the uterus was large and flabby, its walls were thin, and the cervix projected very little into the vagina. Moreover, there were quite distinctly recognised on the lower part of the abdomen some white streaks which in all probability were the remains of the lineæ albicantes. The evidence, when put together, was sufficient to satisfy the jury that the body had been fairly identified as that of Harriet Lane, and the prisoner was convicted.

In the presence of mutilated remains, not only should a careful search be made for anything that will throw light on the mode of death, but anything that might be of assistance in identification of the body should be observed. Any departure from the normal, whether in the shape of the effects of disease, such as caries or necrosis of bone, or ununited fracture, should be noted, or any congenital peculiarity. The shape of the jaw has on more than one occasion been the means of identification of remains which in all probability could not otherwise have been recognised. This was notably the case in the murder of Dr. Parkman. The skull

Means of identification.

was never found at all, but some artificial mineral teeth were discovered in the ashes of a furnace, as well as some melted gold. A dentist was called at the trial, who proved that he had manufactured some artificial teeth for the late professor, consisting of three blocks in combination for each jaw, adjusted with gold fittings. The blocks found amongst the ashes were those he had made for Dr. Parkman, as proved by their correspondence with the trial-plate and mould of the jaw of the deceased, which were still in his possession.

After the total disappearance of the soft parts, the bones will begin to exhibit changes, of which the first is a gradual loss of animal matter, so that the bone becomes lighter. Often, too, the bones become brittle. The rate at which these changes progress depends upon whether the bones are exposed to the air; or buried in the earth, and, if so, in what sort of soil; or in water; as also on the age of the person at death. The ultimate result is that the earthy constituents of the bone undergo disintegration, and the phosphate and carbonate of lime mix themselves with the surrounding earth. The teeth resist much longer than the bones, owing to the extreme hardness of their enamel.

Determination of the age of a skeleton.

The age of a skeleton can be determined with accuracy from the following data.[1] *First year*: The frontal bones are united in half their length; the temporal bone still consists of four separate portions; ossification commences in the lower extremities of the humerus and ulna, in the heads of the femur and humerus, and in the upper cartilage of the tibia. *Eighteen months*: Anterior fontanelle should be nearly closed. *Second year*: Temporal bone united into one piece; the pelvic bones touch each other in the acetabulum;

---

[1] The date mentioned refers to the end, not the commencement; thus, first year signifies end of first year.

ossification takes place in the lower cartilage of the radius and in the tibia and fibula. *Two years and a half*: Ossification takes place in the greater tuberosity of the humerus, in the patella, and in the lower ends of the last four metacarpal bones. *Third year*: Odontoid process united to axis; spinous processes of the vertebræ ossify; ossification takes place in the trochanters. *Fourth year*: Styloid process of temporal bone formed; ossification takes place in the second and third cuneiform bones of the tarsus. *Four years and a half*: Ossification takes place in the lesser tuberosity of the humerus and the upper cartilage of the fibula. *Sixth year*: Frontal suture has disappeared; arches of vertebræ united with the bodies; descending ramus of pubis meets ascending ramus of ischium. *Tenth year*: Ossification in the cartilaginous end of the olecranon. *Twelfth year*: Ossification in the pisiform bone. *Thirteenth year*: The three portions of the os innominatum unite; ossification of the neck of the femur. *Fifteenth year*: Union of the sacral vertebræ; coracoid process united to scapula. *Sixteenth year*: Centre of ossification appears in the acromion; union of olecranon to ulna. *Eighteenth year.* Centre of ossification at sternal end of clavicle. *Eighteenth to twentieth years*: Epiphysis at upper end of femur joined to the shaft, as also those of metacarpus, metatarsus, and phalanges. *Twentieth year*: Upper and lower epiphyses of the fibula and lower epiphysis of femur are united to their shafts. *Twenty-fifth year*: Epiphysis of sternal end of clavicle and of crest of ilium united to the bones. The epiphyses of the bodies of the vertebræ are not united till about the *thirtieth year*. When all the epiphyses are found united, the inference will be that the person has reached adult age. If all the cranial sutures are obliterated, the age of the body may be set down as not less than fifty. The cartilages ossify earlier in women than in men; the cartilages of the ribs become ossified as life advances, as also in old age do the cartilages of the larynx. In old age, too, the diploe of the skull becomes absorbed, leaving the cranial bones thinner and lighter, and the intervertebral discs shrink, which accounts for the diminution in stature of old people, the bodies of the vertebræ falling closer together, and becoming bevelled off in front, thus causing the well-known stoop of old age.

From the femur.

The femur affords some indication as to the age of a body; thus, in early life the neck is directed obliquely, in adults it forms an obtuse angle with the shaft (in the female it is more nearly at a right angle), and in old people the head sinks below the level of the trochanter, and the neck diminishes greatly in length.

From the teeth.

The teeth and the lower jaw afford valuable evidence as to age, the former, mainly in respect to early life. The temporary or milk teeth appear in the following order:—About the *sixth or seventh month* the two middle incisors; by the *ninth month* the two lateral incisors; at the *twelfth month* the first molars; at the *eighteenth month* the canines; by the end of the *second year* the two last molars. The permanent teeth appear between the *sixth and seventh years*, the first molars being the first to be cut; in less than a year, the central incisors should appear; by the *eighth year* the lateral incisors; by the *ninth year* the first bicuspids; a year later the second bicuspids; by the *twelfth year* the canines; from the *twelfth* to the *fourteenth years* the second molars; and between *seventeen* and *twenty-one* the last molars should appear. The teeth make their appearance in the lower jaw sooner than in the upper as a rule. The temporary teeth may be known from the permanent by their smaller size, by the fact that the fangs of the molars diverge at a much wider angle, and by the fact that the enamel at the neck is collected into a sort of ridge, instead of terminating evenly on the fang, as it does in the permanent teeth. Between the ages of six and seven years, in the theoretically healthy child, the jaws should contain the full complement of twenty milk and twenty-eight permanent teeth, more or less developed.

A few remarks suggest themselves about the teeth, which may be borne in mind in examining a skeleton. In the first place, in rickets dentition is often much delayed, and it is not so very uncommon a thing to see a rickety infant of fifteen, or even occasionally of eighteen, months old without a tooth in its head. Another point about rickety children is that they usually cut their teeth irregularly—'across,' as the mothers say—instead of following the order given above. Sometimes children are born with one or more of their incisors already cut. Inherited syphilis certainly, and rickets possibly, cause premature decay of the milk teeth. There are certain defects of the teeth which might be useful in establishing the identity of a person. Such are those cases where the enamel is deficient at the free border of the incisors, or in transverse lines. Inherited syphilis, too, produces certain well-known alterations in the permanent teeth, which might help in the recognition of a body.

The lower jaw will give a rough indication of the age of the person to whom it belonged: thus in the infant, when the teeth have not appeared, or are not fully developed, the alveolar portion of the jaw is imperfect, and the angle between the ramus and the body of the jaw is an obtuse one, *i.e.* greater than a right angle. When the teeth are all present, the alveolar portion of the jaw is of course fully developed, and the ramus is almost, if not

From the lower jaw.

quite, at right angles to the body. In old age, after the teeth have fallen out, the alveolus becomes absorbed, and the jaw once more assumes somewhat of the infantile shape, the angle between the ramus and the body again becoming obtuse.

<u>Peculiarities in the jaw</u> or about the teeth, such as cleft palate, curiously-shaped or supernumerary teeth, often furnish useful means of recognition after death. This was the case in the trial of Webster for the murder of Dr. Parkman already alluded to, and cases have occurred in this country in which the identity has turned upon some question connected with the teeth. Many years ago a body was found near the banks of a stream, and identified, by means of a very peculiar tooth, as that of a man who had disappeared suddenly about ten years before, and a person was arrested on suspicion of having murdered him. When the case came on for trial, the most essential link, viz. the tooth, in the chain of evidence was missing; it had been removed from the jaw. The case broke down at once, for there was no evidence as to the identity of the body, and none to connect the prisoner with it. It is not, however, always necessary that there should be an identification of the body, or indeed that the body should be found at all. In the following case given by Taylor, in which the condition of the teeth was of the greatest value in evidence, the body of the supposed murdered person was never discovered. The case was that of the trial of Elizabeth Ross for the murder of Caroline Walsh. It appeared in evidence that Caroline Walsh, an old Irishwoman, had been repeatedly solicited by the prisoner to come and live with her and her husband, but had always refused, until at length she had consented, and went to the prisoner's house in Goodman's Fields on the evening of August 19, 1831, taking with

Case of Caroline Walsh.

her amongst other things an old basket in which she was accustomed to sell tape and other articles. From that evening all trace of her was lost, and when the prisoner was requested by the relations to account for her disappearance, she made various discrepant statements, finally asserting that the deceased had gone out early on the morning of the day after her arrival and had never come back. The prisoner's son, who was the chief witness for the Crown, testified that the deceased had been suffocated by his mother on the evening of her arrival, and that on the following morning he had seen the dead body of the old woman lying in the cellar, and that later on in the day he had seen his mother go out with something large and heavy in a sack. It so happened that on the evening of August 20 an old woman, answering somewhat to the description of the supposed deceased, was found lying in the street in the immediate neighbourhood, in a completely exhausted condition and in a most squalid state. She stated that her name was Caroline Welsh, and that she was Irish. She was admitted into the London Hospital, her hip being fractured, where she subsequently died. The prisoner Ross, when apprehended, said that this was the woman whom she was accused of having murdered.

The following points were inconsistent, however, with this hypothesis:—The missing woman was 84 years of age, tall, with grey hair and very perfect incisor teeth. Caroline Welsh, the woman who died in the London Hospital was about 60, tall, very dark, and had no front teeth, the alveolar cavities having evidently been obliterated for some time; moreover, her body was emaciated and in a very dirty state, whereas the missing woman was known to have been particularly clean in her habits, and not particularly emaciated. The clothes of the two women were similar; those which had belonged to Caroline Walsh were proved to have been sold by the prisoner, and were produced in court and identified, the clothes of the other woman had been burnt at the hospital. Each had a basket of somewhat similar nature in her possession. The body of the woman who had died at the hospital was exhumed, and the grand-daughters of the missing woman swore that it was not the body of their relative. Even if it had been admitted that they might be mistaken on such a point, the condition of the jaw could not be explained away, as it was clear that the woman who had died in the hospital had not had any teeth for a long time, whilst it was in evidence that up to the time of her disappearance Caroline Walsh had very good incisor teeth. The prisoner was convicted.

<small>Estimation of the height.</small>

In endeavouring to estimate the probable height of a person, from the bones only, the rule is, if the entire skeleton is present, to allow an inch and a half for the soft parts; if the arm alone be given, the rule is to double this and add one foot, which should give the height approximately; from a single long bone it would not be safe to draw any conclusions as to the height of the body. It would be unwise to attempt to express an opinion as to the probable height of a child from the skeleton.

M. de Luca, many years ago, brought some interesting investigations, on the relative length and weight of bones, before the Paris Academy of Sciences. According to him the average stature of an adult man is 5 feet 3 inches, that of a woman one-twentieth less, *i.e.* 5 feet. The head forms one-eighth part of the total height of the body. This is divided into two equal parts immediately below the eyes, while the nostrils are midway between the eyes and the chin. In a vertical section of the body, the centre is at the symphysis pubis; if the arms are raised above the head the umbilicus becomes the centre. The height of a man corresponds to the distance which separates the extremities of the two hands, when the arms are outstretched. The arm may be divided into five parts, the hand representing one part, whilst the forearm occupies two, and the upper arm the remaining two. Therefore, five times the length of the hand will give the length of the arm. In the hand, the carpal and metacarpal bones represent half its length, the first phalanx of the middle finger is equal to one-fourth of the hand, and the two last phalanges of this finger are exactly equal to the first. The last phalanx is divided equally into two parts by the nail. The sole of the foot is a third longer than the palm of the hand, but the back of the foot or instep is of the same length. The observations made on the weights of the bones showed:—1. That the bones of the right side of the body were heavier than those of the left side. 2. That the weight of the bones above the umbilicus was equal to that of those below. 3. That the weight of the bones of the hand was equal to the fifth part of the weight of the whole arm. 4. That the total weight of the hand might be divided into five equal parts, one represented by the carpal, two by the metacarpal bones, and two by the fingers. The weight of the first phalanx was equal to two-thirds of the weight of the entire finger, the other third being represented by the remaining phalanges. 5. That the bones of the hand were half the weight of the bones of the foot. 6. That in the foot the relations were similar. The weight of the bones of the tarsus was double that of the metatarsal bones, and the weight of the toes was divided into three parts, two for the first phalanges and one for the two smaller phalanges.

All congenital peculiarities and evidences of past injury or disease should be noted, as they may prove of use in identification. Sometimes the bones have been burned; in this case they retain their shape, but are very brittle and readily crumble; if burned in the open air they will be white, if in a furnace they will be black or greyish. After being burned they are readily dissolved by hydrochloric acid. If the bones have been reduced to an ash, the finding a large quantity of phosphate of lime in such ash would indicate the fact that bones had been burnt, but would afford no evidence that the bones so burnt had been human and not those of some animal.

It has already been shown that the teeth may be recognised after many years; the hair, too, lasts a long time, and may be found when all else save the bones has disappeared. The methods of examination of hair will be detailed subsequently.

Determination of sex.

The determination of the *sex* is of the first importance in the case of the discovery of a complete skeleton, as until that point has been settled no evidence can be offered as to identity. Up to the age of puberty the skeleton has no distinctive characters in the two sexes; but at the advent of that epoch, important changes take place, leaving behind them alterations which can be recognised as long as the bones retain their shape. Speaking generally, as compared with that of the male, the skeleton of the female is smaller and more slender, and when dried it is lighter. The bones of the female are smoother, not so deeply grooved for the attachments of muscles as those of the male, and the articular surfaces are flatter. The skull of the female is smaller than that of the male; the forehead is lower and narrower, the frontal sinuses are smaller, and the orbits relatively larger, in the woman than in man. The thorax is shorter, smaller, and less prominent in women; the sternum is larger at its upper part, the ribs are shorter, thinner, and flatter, the cartilaginous portions being longer. The clavicles are less bent, the shoulders lower and narrower, the scapulæ are thinner and flatter, the arms shorter, and the carpus and metacarpus smaller and more slender. The neck of the femur in the female forms a less obtuse angle with the shaft than it does in the male; the femur is shorter, more bent, and directed more obliquely inwards; the bones of the leg and feet are smaller and more slender; the bodies of the lumbar vertebræ are higher, and the intervertebral substances thicker, in the female; the foramina, too, are larger.

But it is the pelvis which presents the most distinctive characters; without it, it would be very difficult to express any confident opinion as to sex. The female pelvis is less deep than that of the male, but much wider; the ossa ilii are flatter and more everted, so that the tuberosities of the ischia are more widely separated than in man; the sacrum is wider and turned more backwards; the coccyx is more movable; the symphysis pubis is broader and not so deep; the arch of the pubes is wider and the rami are somewhat everted; the obturator foramen is smaller and more triangular in the female; the acetabula are further apart. The brim of the pelvis is wider, more capacious and more oval than in the male; in the female the greatest diameter is the transverse, in the male the antero-posterior.

The female pelvis.

CHAPTER IV.

SUDDEN DEATH AND PRESUMPTION OF DEATH.

Sudden or unexpected death.

WHEN a person dies suddenly, his health having previously appeared to be good, his death is usually, and ought always to be, the subject of a medico-legal investigation. The causes of sudden, or rather of unexpected death, for the death is not always sudden, are numerous. Poisoning, injury, drowning, lightning, or some other form of violence constitute the majority of such causes. These will be dealt with separately, but there remain a considerable number of cases in which such unexpected death will be due to what are called natural causes—that is, to some disease. Amongst the deaths from natural causes, heart failure, in one form or another, furnishes by far the largest number; indeed, almost all instances of really sudden death can be referred more or less directly to the heart. And of heart diseases, it is especially those associated with disease of the muscular substance of the walls which are likely to give rise to sudden death. Fatty and fibroid degenerations and impaired nutrition, from disease of the coronary arteries, are the most frequent causes of sudden death in people apparently in good health. These changes in the heart walls are apt to take place in persons with valvular disease of the heart, so that sudden death in a person known to have heart

Death commencing at the heart.

SUDDEN DEATH AND PRESUMPTION OF DEATH. 53

disease is far from uncommon. Amongst the less direct causes of sudden death from heart failure may be mentioned hæmorrhage either in the head, thorax, or abdomen, where the death occurs at once from loss of blood. In the brain it is not usually the loss of blood which kills, and, when death is at all sudden in intra-cranial hæmorrhage, it is because of pressure upon, or disturbance of, the medulla oblongata, and may then be regarded as partly due to heart failure and partly to damage done to the respiratory centre. Except in the way just mentioned, sudden death rarely commences at the brain; unexpected death is, of course, very common either from hæmorrhage or embolism or thrombosis, but in these cases death hardly ever occurs with such rapidity as to justify the use of the term 'sudden.' When death commences at the lungs it is not often sudden: the exceptions are cases of obstruction of the glottis by a foreign body, which might almost have been classed under the head of death from violence, and œdema glottidis, which sometimes supervenes very rapidly in adults upon a trifling laryngitis. Almost the only other mode of sudden death in connection with the lungs is that of plugging of one of the pulmonary vessels by an embolus, in which case death is sometimes very sudden. Those rare instances of death from drinking cold water when heated, or from a blow in the pit of the stomach, or under the influence of great mental emotion, are all to be regarded as cases of heart failure. *[At the brain. At the lungs.]*

Presumption of Death.—If a person goes away and is not heard of for seven years, the law is willing to presume, when consulted upon the point, that such a person is dead; but there is no further presumption as to the particular date at which the person died. Anyone wishing to make out that such a person was living at any particular date *[Presumption of death.]*

during that interval will have the onus probandi thrown upon him. Should he succeed, of course that would be tantamount to setting aside the presumption in favour of the person's death, as it could no longer be contended that he had not been heard of for seven years, if his existence at a subsequent date had been proved to the satisfaction of the Court. A person cannot be convicted of bigamy if he or she marries a second time, not having heard anything directly or indirectly of his or her former partner in life for a period of seven years.

Illustrative cases. The following cases given by Taylor may be cited as illustrating the law in respect of this question. A testator died on January 5th, 1861, bequeathing his estate to be divided between his nephews and nieces. One of his nephews had gone to America in 1853 and had written regularly till 1858, when he wrote for the last time from on board an American man-of-war; nothing was known of him after that except that he was struck off the books of the American navy, as absent without leave, on June 16th, 1860. The Court held that his personal representative had not established any claim to a share in the testator's estate, which must be divided amongst those of the nephews and nieces who were proved to have survived the testator. In another instance a policy had been granted on the life of one R. Nutt in 1863. An action was brought in 1874, the question being whether Nutt was then alive or dead; he had been absent from his home for more than seven years. His sister and brother-in-law, who lived where he had formerly resided, gave evidence of his absence, and said that they had not heard anything of him for more than seven years. On cross-examination they admitted that a niece of his had said that when she was in Melbourne in

December 1872, or January 1873, she saw a man whom she believed to be her uncle Nutt, but he was lost in the passing crowd before she could speak to him. No effort appeared to have been made to find him at Melbourne, and the other relatives believed the niece to have been mistaken. The jury expressed a similar opinion. The Judge, however, directed the jury that they 'could not say that the man had not been heard of during the last seven years, when one of his relatives declared she had seen him alive and well within the last three years; and still less could they say that he had never been heard of when all the members of the family admitted they had heard what she had stated;' and 'that the ground for the presumption of death, from a man having been absent for seven years, was entirely removed by the direct evidence that every relative had heard that he was alive.'

On more than one occasion persons have pretended to be drowned by leaving their clothes on the seashore and disappearing, with a view to escape from their debts or to defraud an insurance company. On one such occasion the insurance company paid the premium to the relatives, and the man was recognised a couple of years later in South America. The most elaborate attempt to defraud an insurance company, however, was that made by a Bordeaux merchant, who having insured his life for 100,000 francs was soon afterwards declared a fraudulent bankrupt. He then disappeared, and a month later his wife presented to the office a certificate purporting to be a copy of the register of the death and burial of her husband in England, and claimed the amount for which his life had been insured. A full investigation in this country led to the discovery that the man had passed under various names, had purchased a coffin, procured a certificate from a registrar of

Pretended death.

deaths recording his own death in a circumstantial manner, and that the coffin, filled with a mass of lead, had been buried in a churchyard in Essex.

Priority of death. Presumption of Survivorship.—Questions bearing on this point are constantly arising in the law courts, and do not seem to be always decided on the same lines. The two following cases illustrate this. In the first the question was whether a legatee did or did not survive a testator, so as to take a sum of 4,000*l.* bequeathed to him under the will dated 1858. The testator died in February 1860, the legatee had gone to Australia in 1858, and nothing had been heard of him since January 1859. Seven years having elapsed before the case came into court, he was presumed to be dead, but it remained a moot point whether he had predeceased his father. The Vice-Chancellor said that the law in such cases presumed the continuance of life until the expiration of seven years, when the contrary presumption of death arose; but at what time within that period he died was not a matter of presumption but of evidence, and the onus of proving that death took place at any time within the seven years lay on the persons who claimed a right, to the establishment of which that fact was essential. It was manifest, therefore, that the representatives of the legatee, in order to claim their legacy, must show that he survived the testator; as they were unable to do this, he held that the legacy was never validly given. In another case a solicitor died intestate in 1869. He had three children, two sons and a daughter. The elder son had last been seen in 1868. After the father's death advertisements were inserted in the newspapers for him, but without success. The daughter married and died before the case was heard, her son bringing the action. When she married, she made a settlement that affected any money that came to her by her elder brother's death. Hence the question arose whether the brother was alive when she died, because in that case she could not deal with money over which she had no control. The case came before the Vice-Chancellor in 1876. The first point was, presuming of course that the son was dead, did he die before his father? If so, he would have had no share in the father's estate, which would then have been divided between the two survivors. The Vice-Chancellor decided that the son did survive the father, a decision that seems exactly the reverse of that in the case just recorded, where it was laid down that there was no presumption as to the time at which a person who had been missing seven years had died. It was then asked whether he had died before or after his sister? If the former, his share of the estate would be subject to the marriage settlement of the sister; if the latter, then it would be divided equally amongst his next of kin. The period of the sister's death was known, but not that of the brother. The executors could not prove the date of his death, even if he were dead at all, which they were bound to do. The Vice Chancellor decided that the elder brother having inherited, and by law being presumed to be dead, his inheritance must be divided between the next of kin.

It is, however, in cases of great public calamity that these disputes as to survivorship are most liable to arise, as when a ship is lost at sea, and a large number or the whole of the people on board are drowned, or in the case of a fatal fire. In these cases presumption as to survivorship is based on various considerations; thus, in regard to the persons themselves, the age, sex, strength, and state of health might have some influence, the very young and the very old being naturally presumed to be less capable of withstanding the shock which might be supposed to occur in such a case, than adults. So, too, women are less able to stand great fatigue than men, and the sick than those that are well. The mode of death, too, might furnish some criterion whereon to found a presumption. In cases of drowning, for instance, it would be natural to suppose that a strong man and a good swimmer would keep himself alive longer than a more feeble person who could not swim; but, as will be seen directly, these presumptions, founded on common sense or on physiological knowledge, are not allowed much weight in a court of law. Any evidence, however slender, as to the fact of one person having been seen alive later than another in the case of a shipwreck, will override a probability as to which would have been most likely to survive. As a rule, when a ship is lost at sea, the law assumes that all hands were lost at the same time, unless there is direct evidence to the contrary, so that in the case of a married couple, one of whom was possessed of some property, the law will assume that neither survived the other, and the property would remain vested in the person who had possessed it, and his heirs. The principle upon which the law acts in these cases is that a right to property ought not to be obtained or transferred upon mere presumption, however strong or likely. Thus, in the case of Underwood *v.* Wing, the Court refused to make any presumption as to priority of death. The case was one of extreme importance, and was carried through all the law courts and to the House of Lords, but invariably with the same result. The facts, as recorded by Taylor, were simple. Mr. and Mrs. Underwood, being about to go to Australia with their children, respectively made their wills, each

Underwood and Wing.

giving to the other absolutely the whole of their properties, and by each such will they declared that if the one to whom the same was given should die in the lifetime of the donor, the property should be divided among their three children on their attaining majority, and that in case all their children died under the age of twenty-one, they then directed that their property should go to their mutual friend Mr. Wing. The ship in which they sailed was wrecked off Beachy Head, and Mr. and Mrs. Underwood with all their children were drowned. As the children were all under age, the only question that arose was whether Mr. and Mrs. Underwood perished at the same moment, because this contingency was not provided for in their wills, and if it occurred the next of kin would inherit. So far as evidence went, it was proved that they were swept off the deck at the same time, and when last seen Mr. Underwood was supporting his wife in the water and the two boys were clinging on to her. Mr. Underwood was a big, powerful man and a good swimmer, Mrs. Underwood was a slight, delicate woman, and therefore, from a medical point of view, there was every reason for believing that he would have survived his wife for a short time at any rate, whilst, physiologically speaking, it was almost impossible that they should have died at the same moment. In every instance, however, the Court decided against the defendant Wing, and in favour of the next of kin. The onus of proof was thrown on Wing, his contention being that the husband and wife could not have died at the same instant of time. This negative proposition could not, of course, be proved by direct evidence; it was a matter of medical inference; and the law declares that, in the absence of evidence, the property shall go in the same way as if the parties had died intestate.

As regards the presumption in other forms of death, in asphyxia women are said to have a better chance of surviving than men; heat is better borne at the extremes of life than in middle age; with cold the reverse holds good. In cases of starvation the young succumb first, and the aged endure the longest; corpulent persons withstand hunger better than the lean. When mother and child have died in childbirth, the presumption is that the mother lived the longest.

CHAPTER V.

DEATH FROM COLD, HEAT AND STARVATION.

DEATH from cold is in this country happily rare, and when it does occur is probably partly associated with starvation. The first effect of prolonged exposure to cold, beyond the mere bodily suffering caused by it, is often a tendency to lethargy, which, unless vigorously resisted, will prove the forerunner of death, the person sinking into a state of torpor from which he never awakens. When exposed to intense cold the skin, at first purple, slowly becomes white, the muscles stiffen gradually, so that their contraction becomes increasingly difficult; this applies especially to the muscles of the face and extremities. Before loss of sensibility supervenes there may be giddiness and dimness of sight. Sometimes convulsions have been noticed. Death from cold is encouraged by conditions producing bodily exhaustion, especially of the nervous system, no matter how caused. Old age, disease of any sort, and habits of intemperance render people more susceptible to cold, and therefore less able to withstand it. New-born infants readily succumb to cold.

Death from cold.

Cases of murder by cold are very rare, and it is to be hoped that the following case, given by Taylor, will long be unique:—A man and his wife residing at Lyons were tried for the murder of their daughter, aged 11, under the following circumstances: On December 28th, at a time when there was a severe degree of cold, the woman compelled the deceased to get out of bed and place herself in a bath of ice-cold water. The child cried and endeavoured to escape from the bath, but she was by violence compelled to remain in the water. She soon complained of exhaustion and dimness of sight; the prisoner then threw a pail of iced water on her head, soon after which the child died. Death was properly ascribed to the results of this maltreatment, and the woman was convicted. Such a case could only be proved by the circumstantial evidence, for, as will be seen directly, the post-mortem appearances are not sufficiently characteristic to be decisive of death from cold.

Ogston's conclusions. Dr. Ogston, who had the opportunity of examining the bodies of sixteen people who had died from the effects of cold, states that he met with several points of agreement, from which he was led to conclude that, when they were all present in the same case, and when there was no other obvious cause of death, they rendered it highly probable that death had been due to cold. The appearances to which he refers were: 1. An arterial hue of the blood generally, except when viewed in mass within the heart, the presence of this colouration not having been noted in two instances. 2. An unusual accumulation of blood—as in Quemalz and Cappel's cases—on both sides of the heart, and in the larger blood-vessels of the chest, arterial and venous. 3. Pallor of the general surface of the body, and anæmia of the viscera most largely supplied with blood. The only exceptions to this were moderate congestion of the brain in three of the cases, and of the liver in seven of them. 4. Irregular and diffused dusky-red patches on limited portions of the exterior of the bodies, encountered in non-dependent parts, these patches contrasting forcibly with the pallor of the skin and general surface. He did not, however, find these appearances so uniformly present in children as in adults; the arterial hue of the blood was absent in one child, pallor of the surface was absent in another; anæmia of the internal viscera was absent in all but one of the cases. Tidy points out that experience has shown that the very young and the very old have limited powers of heat production, and they therefore bear extreme cold very badly. He also calls attention to the fact that, in some respects, the effects of extreme heat and extreme cold are very similar. For instance, they both produce vesications.

Death may follow exposure to great heat without necessarily any exposure to the rays of the sun. Those employed in engine-rooms and others exposed to great artificial heat sometimes die very suddenly. In such cases the brain is usually found congested, and sometimes an effusion of serum into the ventricles is present. The drier the atmosphere the greater is the amount of heat that can be borne without danger. Dr. H. C. Wood, who has had a large experience of deaths from sunstroke or insolation, divides the cases of death from extreme heat into three classes. In the first there is heat exhaustion with collapse, accompanied by rapid feeble pulse and a cool moist skin, with a tendency to syncope. In the second there is intense heat of the skin, which is dry, with a rapid feeble pulse and delirium. In the third class, which is rarer than the others, there is meningitis or phrenitis—true *coup de soleil*. In these cases the temperature sometimes rises very high, even to 110° F., or higher, sometimes rising a little after death. The breathing is laboured and often stertorous; the face is congested. Subsultus tendinum, spasms, or general convulsions are often noticed. Dr. Wood

Death from heat.

found the heart firmly contracted in all the bodies which he examined. Besides this there is nothing important to note as to the morbid appearances; rigor mortis sets in early, and so does putrefaction.

Death from starvation. Starvation is of two kinds, acute and chronic, according as it is complete or partial. The former differs from the latter chiefly in the fact that, as its name implies, the symptoms do not last so long, the person seldom living more than ten days when the abstinence from food has been absolute; whereas in chronic starvation the time of death depends entirely on the degree of completeness of the deprivation of food.

Starvation may be accidental, suicidal, homicidal, or pretended. Accidental cases are those where the person is precluded by his surroundings from obtaining any nourishment, as in the case of miners shut up in a mine or in cases of shipwreck. It also may be considered to include those cases in which the starvation is the result of disease, *e.g.* in cases of carcinoma of the œsophagus or of the stomach, in which there is practically a state of involuntary starvation. Suicidal cases are nearly always, if not always, an evidence of insanity, refusal to take food being a very common symptom amongst the insane. Homicidal cases are fortunately not very common; amongst them may be included those instances where infants have been wilfully starved to death by their parents or those in charge, as the most convenient way of getting rid of them without attracting too much attention. Instances of pretended starvation have occasionally occurred in this country, as will be seen later on, where designing persons have traded on the credulity of the wonder-loving public.

Chronic starvation. The following list of symptoms refers more to cases of chronic than of acute starvation. Abdominal pain, relieved by pressure, is early complained of; the countenance becomes pale, the eyes glistening. There is intolerable thirst; the mouth is dry, and the saliva becomes thick and viscid. The patient steadily emaciates, rapidly at first, more slowly after a time. The urine is secreted in very small quantity, and is therefore very concentrated; the fæces are scanty, dry, and solid. The pulse is usually at first somewhat quickened, afterwards becoming slower; the temperature may not be much affected until towards the end, when it sinks. The mental faculties gradually become obscured in proportion to the bodily weakness; sometimes there is delirium.

Experiments by M. Chossat on the loss of weight in animals during starvation led to the conclusion that death supervenes when the animal has lost 40 per cent. of its weight. This loss of weight is of course differently distributed amongst the various tissues and organs of the body. Thus the fat loses 93 per cent., the blood 75 per cent., the spleen 71 per cent., while the nervous system loses only 2 per cent. Hippocrates says on this subject: 'The old bear the want of nourishment best; they who have attained the middle period of life the next in degree; those who have just arrived at puberty are still less able to bear it, but of all ages childhood is the least capable of sustaining hunger, and of children the more lively are the least capable.' Other things being equal, fat people will bear privation the longest.

After death in a case of chronic starvation the body will be found exceedingly emaciated and quite devoid of any subcutaneous fat, the skin being dry and shrivelled. The muscles are wasted and have lost their fat; the same applies to the internal viscera, the heart and kidneys especially having lost that surrounding of fat which is usual to them in a state of health. The alimentary canal *Post-mortem appearances.*

is empty, and the stomach and intestines are usually more or less contracted; the thinness and transparency of the walls of the small intestine are points on which a good deal of stress has been laid. The liver and spleen do not contain their usual amount of blood; the gall-bladder is distended with thin bile, and the omentum is devoid of fat. The bladder is empty and contracted. A peculiar and unpleasant odour will be perceived, in most cases it may even be recognised during life.

In dealing with a person suffering from the effects of prolonged starvation, it is important to remember that the administration of food must be very gradual and in a form easily digested. Those who disbelieve in the genuineness of the celebrated forty days' fast (so-called) of Dr. Tanner, point to the fact that he is said to have eaten a beefsteak at the termination of it as supporting their view.

Pretended starvation. Of pretended fasting a good instance was afforded in the commencement of this century by Ann Moore, the celebrated fasting woman of Tetbury, who gave out that she had taken no food for six years. A first committee having failed, a second committee was appointed to watch her, and at the end of nine days she confessed the imposture.

This happened at a time when physiology was comparatively in its infancy. For the public credulity in the next case, which occurred in 1869 *The Welsh fasting girl.* in the person of Sarah Jacobs, aged 12 years, there was no such excuse. After some acute illness the child fell into an hysterical state and would not take her food; the parents said they would not press her if she did not wish it, and to this they adhered throughout, saying they would give her food if she asked for it, but not otherwise. The child took to her bed, and people used to come from the immediate neighbourhood and bring her presents of flowers and money, the parents always refusing to accept any money themselves though encouraging visitors to give it to the child. When this had been going on for about eighteen months, in an evil day for the child, the matter got into the daily papers, and soon after a committee was appointed, including four medical men. Four nurses from one of the London hospitals were then invited, with the consent of the parents, to come down and watch the child. The instructions they received were merely to watch, and report if any food was given to the child; they were not in any way to prevent her having food, and if she asked for it they were instructed to give it to her. The doctors were merely to look on. The result was that at the end of eight days' watching, as might naturally have been expected, the child died; the parents declaring to the last that they had never given the child anything for two years, and that they would not unless she asked for it. At the autopsy there was a good layer of subcutaneous fat over the abdomen, the alimentary canal was almost empty, save a few scybala in the large intestine; the gallbladder was empty and the urinary bladder distended. In short, the appearances were those of acute starvation, and quite inconsistent with the idea of her having been without food for anything approaching the length of time which the parents asserted to be the case. The parents were indicted for the manslaughter of their child, and were convicted of causing her death by criminal negligence.

One other case, of homicidal starvation, requires mention. In 1877 a man named Staunton was tried with others for the murder of his wife and convicted. The body was extremely emaciated, and in a most filthy state, and the evidence left no doubt that the deceased had been most brutally ill-treated; but the medical men who made the examination were not agreed as to the absence of

Homicidal starvation.

F 2

disease, one gentleman describing what he believed to be a mass of tubercle on the surface of one hemisphere of the brain. After the conviction several members of the profession pointed out that the presence of this tubercular disease would account for many of the symptoms, and was in itself a sufficient cause of death. In consequence of these representations the sentences were commuted.

Asphyxia.

Symptoms.

- Abortive effort to breathe
- Cyanosis – veins turgid
- Pulse feeble
} *Forbes 3 stages* — Dyspnœa

- Convulsive twitchings of limbs
- Blood-stained foam fr. mouth
- Tongue semi-protruded
} II / Convulsion

- Consciousness early lost
} III / Exhaustion

Death from drowning may be due to
I. asphyxia pure & simple or
II. Syncope
III. Exhaustion
IV. Cramp
V. apoplexy
VI. concussion
} may be asphyxia + these causes

CHAPTER VI.

DEATH FROM ASPHYXIA.

ASPHYXIA really means pulselessness, but it has so long been used in the sense of want of breath, that any attempt to give this up would probably lead to confusion; moreover, the physiologists have attached a totally different meaning to the word 'apnœa,' which would be the only one that could be substituted for it; apnœa, according to them, being the condition in which the blood is too highly arterialised—*i.e.* contains too large a proportion of oxygen, and consequently fails to stimulate the respiratory centre in the medulla oblongata, the result being that respiration does not take place. By asphyxia then is meant a <u>condition in which a due supply of air is not obtained, and the aeration of the blood is not performed.</u> A state of dyspnœa ensues, the respiratory movements become much hurried, and unconsciousness follows with more or less speedy death.

Definition.

Asphyxia is the cause of death in drowning, hanging (not the judicial form), strangulation and suffocation. The symptoms preceding death in such cases are at first abortive efforts to breathe, accompanied by a gradually deepening cyanosis, with turgidity of the superficial veins; the pulse becomes more and more feeble; convulsive twitchings of the limbs make their appearance, and a

Symptoms.

blood-stained foam exudes from the mouth; the tongue is often semi-protruded, and its base much congested. Consciousness is lost early, but the mind is unusually active before this stage is reached. Foster distinguishes three stages: 1. A stage of dyspnœa characterised by an increase of the respiratory movements, both of inspiration and expiration. 2. A convulsive stage characterised by the dominance of the expiratory efforts, and culminating in general convulsions. 3. A stage of exhaustion in which lingering and long-drawn inspirations gradually die out. Death is rapid in all cases of asphyxia. The post-mortem appearances common to all cases of asphyxia are the following: General lividity of the body-surface with variously-sized extravasations of blood into the skin, usually small. The dependent parts of the body are often very deeply stained. On opening the body the right side of the heart, and the venous system generally, will be found engorged with black fluid blood, the left side of the heart and the arteries appearing empty. The mucous membrane of the trachea and bronchi will be deeply congested; in young subjects the lungs may be found to have emptied themselves of blood owing to the violence of the respiratory efforts, and a little emphysema of the lungs from the same cause is common. In some cases, especially in suffocation, ecchymoses will be found on the pleuræ and other serous membranes. The membranes and sinuses of the brain share in the general venous engorgement. The pleuræ, pericardium, &c., usually contain an excess of serum. The abdominal viscera, especially the spleen and liver, are engorged with dark blood. Rigor mortis is late in making its appearance.

Death from Drowning.—On the discovery of a dead body in the water the first point to be determined is whether death has been due to drowning

<small>Post-mortem appearances.</small>

<small>Drowning.</small>

or not. If not, then the cause of death must be sought, and the question whether death took place in the water, or not, must be answered. Supposing the death to have been due to drowning, it will be important to see what evidence there is as to whether this was suicidal, accidental, or homicidal. Lastly, the time that the body has been in the water must, as far as possible, be determined.

As regards the actual phenomena of drowning, the following graphic description, taken from Devergie, is given by Ogston: The individual sinks in the water to a depth greater or less, according to the height of his fall; then rises again to its surface, under the influence of his specific gravity, rendered less considerable by the air retained in his clothes, and by the position which the body assumes by the effect of his instinctive actions, which have for their object the presentation of a greater surface to the fluid. Then one of two things happens: either the individual can swim, and in this case he instinctively pushes along the surface until he is fatigued, when he comes to the same position as the person who cannot swim; or he finds himself from the outset in this latter predicament, and then he executes irregular movements of the arms and legs, seizes everything within his reach, clutches at the bottom, lays hold on bodies in motion as motionless ones, but, inasmuch as his motions are irregular, he appears and disappears successively from the surface of the water. It has been observed that, at the moment the head comes to the surface, air and water are inspired; the latter is partly swallowed, partly rejected by an involuntary fit of coughing, arising from the contact of the water with the larynx, the liquid reaching this organ at the same time with the air. These efforts, however, have caused the expulsion of the inspired air, and the desire to breathe makes itself imperatively felt.

Mode of death.

If the individual has been able to reach the surface of the water he profits by his contact with the air to satisfy this desire for breath; but, as the head is only partly above or out of the water, he draws in a fresh quantity of air and water, and the cough returns. In a short time the person only floats beneath the surface; he again feels the desire to breathe; he opens his mouth, and water alone enters it; it is expelled from the trachea mingled with air or perhaps swallowed; and a quantity of it, between one and two pounds, may thus reach the stomach. A little water always enters the trachea, to form the foam which is found when the body is examined early enough. During the whole of these attempts to keep up the respiration, an afflux of blood is taking place towards the brain, which explains the bloody points, and even the gorged state sometimes discovered in that organ, which is never found in mere asphyxia. Finally, the voluntary movements cease, the asphyxia is complete; the individual sinks to the bottom of the water, while at the same time bubbles of air escape from the mouth, on account of the collapse of the walls of the chest and of the diaphragm, under the influence of their elasticity.

As regards the sensations of the person, a sort of delirium is described as coming on, followed by a ringing sensation in the ears, after which the person loses all consciousness; in cases of recovery there is no recollection of any suffering during the process of asphyxia. Vomiting often takes place, and may assist in bringing about the fatal issue if, as sometimes happens, the vomited matters pass into the trachea. This is so real a danger, that it has been suggested that, in cases of apparent death from drowning, tracheotomy should be one of the earliest steps taken on account of the possibility of the larynx being blocked by food.

DEATH FROM ASPHYXIA.

Causes of death.

Death is not always due directly or solely to asphyxia, and Tidy gives the following list of secondary or supernumerary causes of death :—1. Syncope (neuro-paralysis). This may result from fright, from drunkenness, from hysteria or catalepsy; but unless the syncope were so complete that death had practically taken place before the person reached the water, there would be some signs of asphyxia. 2. Exhaustion. As in the case of a good swimmer. 3. Concussion. From coming into contact with the bed of the river, rocks, posts, &c. 4. Apoplexy. Coming on either during the violent struggles in the water, or just before the person jumps into the water. 5. Cramp. The muscles of the extremities being first tetanised, the spasm extends to the muscles of respiration and perhaps to the heart itself. As additional causes he mentions shock and an epileptic seizure.

A very short period of complete submersion suffices to destroy life; but it is not desirable to make any positive statements as to how soon death will occur, or as to when it would be hopeless to attempt resuscitation. A case has been recorded in which recovery took place after twenty minutes' submersion in consequence of prompt treatment. Divers can endure a longer period under water than ordinary people, because, according to physiologists, their respiratory centre gets used to the irritation of venous blood, and is, therefore, not so speedily affected by it. Even they, however, cannot stay under water much more than a minute with comfort or safety.

Some years ago a Committee of the Royal Medical and Chirurgical Society was appointed to investigate this subject, especially with reference to treatment, and they made some very interesting observations on dogs. They found that four minutes' complete submersion in water killed dogs, though the heart might continue to beat after removal from the water. Amongst other experiments the following may be quoted here: Two dogs of the same size were submerged at the same moment, but one had his trachea plugged and the other had not. At the end of two minutes they were removed from the water—the one that had his trachea plugged recovered at once, the other one died. In other instances animals with the trachea plugged were submerged for four minutes, and recovered perfectly when taken out of the water. From this it was inferred that one main cause of death in drowning was the distension of the lungs by water which prevented their collapsing; and it afforded a good explanation of the fact that a person, who had fallen into the water in a state of syncope, was not in such danger of being drowned as one falling in whilst perfectly conscious.

Treatment. As regards the treatment of the apparently drowned, the following are the directions of the Royal Humane Society:—1. Remove all clothing from the neck and chest. 2. Wipe the body dry and cover it with dry clothes. 3. Clear the nostrils, mouth, and throat of all mucous froth or substances likely to interfere with free respiration, pull the tongue forward, and keep it in this position, so that it may not fall back and cover the entrance to the glottis. 4. Place the body at full length with the face downwards; the forehead resting on one arm, this will allow any fluids to flow out of the mouth readily. 5. Ammonia, aromatic vinegar, snuff, or other stimulants may be cautiously applied to the nostrils. 6. If respiration be not quickly re-established spontaneously, then the body should be placed on its back, with the head slightly raised, and artificial respiration performed.

Silvester's method. Silvester's method is the one usually adopted in this country. The patient is placed upon his

DEATH FROM ASPHYXIA.

back; the person who is to perform the artificial respiration then grasps the arms near the elbows and slowly brings the arms up over the head, keeping them fully extended for two or three seconds; he then brings them down to the side and compresses the thorax laterally at the same time; this manœuvre is repeated about fourteen times a minute. The chief thing to be borne in mind is not to be in too great a hurry, but gradually and regularly to introduce air into the chest and squeeze it out again, imitating as nearly as may be the natural rhythm of respiration. Another method which has not yet found much favour in this country is that of an American doctor named Howard. His directions are the following :—' 1. Instantly turn the patient, face downwards, with a large firm roll of clothing under the stomach and chest. Press with your weight two or three times, for four or five seconds each time, upon the patient's back, so that the water is pressed out of the stomach and lungs, and drains feebly downwards out of the mouth. Then, 2, quickly turn the patient's face upwards, the roll of clothing being now put under his back just below the shoulder-blades, the head hanging back as low as possible. Place the patient's hands together above his head. Kneel with patient's hips between your knees. Fix your elbows against your hips. Now grasping the lower part of the patient's chest, squeeze the two sides together, pressing forward with all your weight for about three seconds, until your mouth is nearly over the mouth of the patient; then with a push *suddenly* jerk yourself backwards. Rest about three seconds, then begin again. Repeat these bellows-blowing movements, so that the air may be sucked into the lungs about eight or ten times a minute. Remember, the above directions must be used on the spot directly the patient is taken from the water. A moment's delay, and

Howard's direct method.

success may be hopeless. Do not stop these movements under an hour unless the patient breathes.'

With these measures for the restoration of the breathing others may be usefully employed, in the shape of warmth to the extremities, and the use of stimulating friction; the application of heat is especially recommended to the region of the heart, to the loins, the soles of the feet and the palms of the hands. When spontaneous breathing has been established, hot fluids may be administered by the mouth, and the person should be put to bed and kept warm.

Favourable signs. Amongst the signs, which indicate the probability of a favourable termination, may be mentioned slight flushing of the face with convulsive twitching of the muscles, warmth of the skin, and convulsive movements of the limbs. The unfavourable signs are coldness and pallor, dilatation of the pupils, stiffness of the lower jaw with the presence of frothy mucus about the lips. The method of performing artificial respiration by means of faradising the phrenic nerves may be mentioned, though, of course, in the vast majority of cases it would be quite impossible, owing to the absence of apparatus. It is performed by placing one pole over the nerve in the neck on each side, it being accessible towards the lower part of the neck where it lies on the anterior scalenus. Connection should not be made for more than a few seconds at a time, the object being to induce contraction of the diaphragm.

The fact that a case has been recorded in which an adult male, said to have been submerged for fourteen minutes, recovered after eight and a half hours' assiduous treatment, should be very encouraging to those called upon to treat an apparently hopeless case. It must never be forgotten, however, that recovery does not always take place, even when the breathing has been satisfactorily re-established;

death occurring, sometimes on the same day from shock, sometimes later from other causes. The two points to be borne in mind are, first, not to neglect treatment because a case appears hopeless—of course it is not intended to advise treatment when decomposition has evidently set in—and in the second place, if success is to be attained, the treatment must be prompt and assiduous.

The post-mortem appearances will depend upon how long the body has been in the water, and also how long a time has elapsed since it was taken out. *Post-mortem appearances.*

As regards the external appearances, it may be as well to state, at the outset, that the death of a person by drowning cannot be established by the external appearances alone. Still the changes that are found in the skin, &c. have a certain value as pointing to a presumption of drowning, and they must, therefore, be briefly considered.

The body will be cold and the skin pallid, though, according to Ogston, if the body has been in the water for about sixteen hours a rosy tint of skin is commonly seen; the condition known as the *cutis anserina*, or goose-skin, will generally be present. This has been looked upon as a sign of much value, indicating that death was due to drowning. Though it has not quite this meaning, still it is a sign of some importance; for it shows that if the person was not living when he fell into the water, at any rate the body was warm when this happened, and it might well be asked. Why should a body, whilst still warm, be placed in the water if death had really been due to natural causes? The face is pale, calm, with a placid expression. Often there is froth about the mouth and lips—the value of this sign will be seen when the internal appearances are spoken of; the tongue is said by some authorities to be swollen and pushed against the teeth so as to be indented by them; but several eminent writers, including Casper, have not found this to be the case. It depends upon the length of time that the body has been in the water whether the limbs are found rigid or relaxed, but rigor mortis usually sets in early, and the limbs may be found as in the attitude of struggling. Abrasions will often be found on the surface of the body, but too much stress should not be laid upon them, as it is obvious that they might be caused by the body being washed against rocks, &c. after death.

The presence in the hands of gravel, sand, mud, or other substances belonging to the bed of the stream or river where the body of the person was found, would point strongly to the presumption of death by drowning; as also would the fact of a person being found with a tight grasp of a rope or of another body; the absence, however, of any of these conditions should not be allowed to stand as any argument against the hypothesis of death from drowning. If the body has been in the water a few days, the skin of the palms and soles becomes white and sodden. Another sign described by some writers is that known as the cholera hand, in which the skin seems unusually translucent, the hand appearing bluish through it. Some stress is laid by Casper on retraction of the penis as a sign of death by drowning, but Ogston says that he only met with it six times, and that on twenty-two occasions he found, on the other hand, a state of semi-erection.

Water in the stomach.

As regards the internal appearances, the two most important signs are, first, the presence of water in the stomach, and next, the presence of water in the air passages. As to the first of these, the presence of water in the stomach; it has been found by experiments on animals that in an animal put into the water after it was dead, or when it had been stunned or was in a state of syncope, no water was found in the stomach; and that the amount found in the stomach of an animal drowned by being put into the water alive, and not in any of the above-mentioned conditions, was greater, if the animal were allowed to come up to the surface several times, than if it were submerged once for all. In fact, in animals kept entirely under water from the first, there was very often no water in the stomach at all. If putrefaction had set in, the same value would not attach to the discovery of water in the stomach. The absence of water from the stomach would not be a proof that death had not been due to drowning.

The presence of water, mucous froth, mud, sand, gravel, &c. in the trachea and bronchi affords very strong evidence of death by drowning; it would indeed be difficult to explain the presence of some of the latter on any hypothesis, which did not presuppose that the person was living when he went into the water. Frothy mucus, as has already been mentioned, may be found about the mouth; it is not always found in the trachea when death has been due to drowning, though very generally so, and is often mixed with a little blood. It depends on the violence of the efforts made to breathe, and on the length of time that the body has been submerged; unless the examination be made within one or two hours after death, the frothy condition may have all passed off. Water, mud, sand, &c. may be found to have penetrated into the smallest bronchi and air vesicles, and in consequence the lungs seem more or less distended, and do not collapse when the thorax is opened; they are also generally somewhat engorged with blood. The only other sign with regard to the respiratory tract that need be noticed is the upright condition of the epiglottis that has been described; it does not, however, appear that this is by any means constant.

Water, &c. in the lungs.

As regards the other appearances in the bodies of the drowned, the venous system is usually engorged, and the blood is often, though not always, fluid; the heart most commonly will be found with the right side distended and the left empty, but this is by no means always the case, and both sides may be nearly empty or nearly full. The brain is not generally congested, though the sinuses are often full of blood. If the drowning took place whilst digestion was going on, there will be more or less redness of the mucous membrane of the stomach; or if some little time has elapsed since death, there may be a purple appearance. The abdomen may be much distended from inflation of the intestines with gas. The condition of the bladder varies.

Was death due to drowning? The question as to whether drowning was the cause of death or not is one that arises every time that a dead body is found in the water; and it must be borne in mind that, besides suicide, a person may fall into the water and be drowned in an epileptic fit, or a seizure of apoplexy, or during fainting; or that he may have been put to death and been thrown into the water afterwards. As regards any of the first causes, viz. the person falling into the water in a state of unconsciousness and being drowned, all the signs of drowning that have just been under consideration could not be expected; there would be no evidence of struggling, no water would be swallowed, there would be no froth in the trachea and bronchi; and the signs would be partly those of asphyxia and partly those of syncope. If the person had been suffocated, and his body then thrown into the water, it is difficult to see how the fact could be proved in the absence of any marks of violence about the neck. If evidence of disease likely to have caused death, or marks of external violence sufficient to account for death, were found, and were such as to render it improbable that a person could have inflicted them on his own person, and then thrown himself into the water; then all the circumstances of the case ought of necessity to be investigated with minuteness and care.

The difficulties encountered in some such cases are very great, as the following instance will show. On January 1st, 1884, the dead body of a young man was found in the water at a reservoir in the outskirts of London. There was a handkerchief round his neck, tied with three knots. A little distance from the edge of the water there were marks of footsteps, which, in the opinion of the medical men who examined them, indicated that a struggle had taken place there. An earring picked up on this spot seemed to confirm this theory. The young man's watch was missing. A friend of the deceased, who had parted from him after midnight, gave evidence that he was then in good health and spirits, and that he knew of no reason why he should commit suicide. The place where his body was found was nowhere near where he lived; nor after parting from his friend would he have had to pass the reservoir to reach his lodgings.

The post-mortem examination was conducted by Mr. Bond, who thus described the results of his investigation: 'The body was that of a well-grown, fairly muscular man. Rigor mortis was still present, and there was no sign of decomposition. The face was red and suffused, and the mucous membrane of the lips was purple. There was a slight abrasion of the upper lip, as if it had been pressed against the teeth. Frothy mucus, slightly tinged with blood, issued from the nose. There was a small lacerated wound on the tip of the nose, evidently done during life or immediately after death. There was no fracture of the nasal

G

bones. There was an abrasion on the lobe of the left ear, and this abrasion also appeared as if made during life. The eyelids were closed, and there was no protrusion of the eyeballs. There was no congestion of the mucous membrane of the eyes, which were natural. The legs were partly flexed at the knee-joints, and the hands were partly clenched. Around the front of the neck there was a white mark, not depressed, with a red rim of congestion above and below, half an inch wide. It extended three-and-a-half inches back on the right side and four on the left. On the left side there was also a depression at the extremity of the mark, as if by a knot, and red in the centre. The line on the neck was on a level with the lower part of the larynx, and it sloped slightly upwards at the back. The cellular tissue under the white line was bloodless and of a light colour, with a rim of congestion above and below the white mark. This condition of the neck indicated, in my opinion, that pressure had been made during life, or immediately after death, by a soft thick ligature like a handkerchief. There were no other external marks of injury, even on the chest. The skin of the chest was somewhat mottled or spotted. The scalp was exceedingly vascular, but there was no suffused blood under it, and no signs of a blow on the head. The membranes of the brain were intensely congested, and especially so the veins. On removing the skull-cap, eight ounces of black fluid blood escaped. There was considerable arterial congestion of the brain, but no fluid in the ventricles. There was no injury or fracture of the skull, and the substance of the brain was healthy. The nose and mouth were full of frothy mucus. The back of the tongue was slightly congested, and the epiglottis was of a brick-red colour from congestion. The mucous membrane of the larynx

DEATH FROM ASPHYXIA. 83

was congested throughout. The surface of the lungs was much mottled, and very crepitant on pressure. There were light-coloured patches scattered all over the surface of the lungs. These patches disappeared on pressure with the finger. They were caused by air-bubbles from ruptured air-cells beneath the pleura. This condition is very marked, and is found after death from strangulation. The lungs were much congested throughout, and full of frothy mucus. There was no water in the lungs, and they were different from the lungs of persons who had been drowned. There was slight congestion of the bronchial mucous membrane, and extravasation of blood into the lung-substance; and there was no lung disease. There was about a drachm of clear fluid in the pericardium. The left ventricle was contracted, and nearly empty; the right relaxed, and nearly empty; but the pulmonary vessels were gorged with blood. The mitral valve was thickened with adherent nodules of old lymph; but the heart was otherwise empty. The stomach contained about eleven fluid ounces of semi-transparent liquid, mixed with nearly digested food. There were no weeds or mud in the stomach. The stomach-contents smelt strongly of alcohol. The stomach itself was quite healthy. The kidneys and other abdominal organs were healthy.' Mr. Bond expressed a very strong opinion that death was due to strangulation or suffocation, and not to drowning, and further, that it could not have been the result of suicide. It appears, however, that the police subsequently obtained information which led them to decide that the case was one of suicide, and that no murder had been committed; though the exact grounds upon which that conclusion was based were never made public.

The specific gravity of the human body is but Floating of
drowned
bodies.

slightly greater than that of the water, and very little movement is required to prevent the body from sinking. A fat body will float more readily than a lean one, and a body with large bones sinks more easily than one with small bones. The bodies of infants, being generally provided with a good superficial layer of fat, float readily, and in some cases of infanticide the body never sinks at all.

The clothes, too, may make a difference; at first they may buoy a person up, afterwards they will help to weigh him down. The period at which a body will rise after being drowned depends on: 1. Specific gravity. 2. The nature of the water, whether salt or fresh. 3. The access of heat and air in producing decomposition. Putting aside the cases where it never sinks at all, a body does not generally rise until putrefaction has commenced. It is impossible to fix any date as to the time at which a body may rise, or to lay down a hard-and-fast law that it cannot rise sooner.

Illustrative cases. An important case occurred in America some years ago, where a question arose as to whether a body could rise within three days of death by drowning. In this instance the body had been attacked by fish and crabs; this fact, in the opinion of some of the witnesses, pointed to a much longer period since the death than was supposed. The evidence of identification was very slender, and there were some suspicious circumstances in the case; thus, the person alleged to be drowned had been insured for a rather large sum of money, and the insurance company, who were the defendants in the trial, believed that he was not dead at all, and that the whole thing was a ruse to get money. The jury, however, found a verdict against them, declaring that the body that had been found was that of the missing man, and thus affirming that a body could rise in three days. This fact has been established by the occurrence of several indubitable cases in this country (where there could be no possibility of mistake), of bodies being found floating on the water within twelve hours of death. One of the most important and best known of these may be briefly alluded to here: A barrister, who afterwards rose to great judicial eminence, was indicted for the murder of a certain Mrs. S——. They had been staying at a country house, and one morning Mrs. S—— was missing. Thirteen hours later her body was found floating in a stream; it was buried, but six weeks after was exhumed and examined. No water was found in the stomach or lungs, which it is stated were not putrefied. At the trial, in reference to the fact of the body being found floating, a sailor declared that this was an evidence that the deceased had not died from drowning, but had been thrown into the water after death; and he explained that for this reason a weight was always attached to a dead body when it was buried at sea. The medical evidence revealed nearly as much ignorance on the part of the witnesses. Six medical men testified that when a person was drowned, water was invariably taken into the stomach and lungs; and as none was found in this instance, they were of opinion that deceased came to her death by other means than drowning—in fact that, as alleged in the indictment, she had been murdered by the prisoner, and her dead body afterwards thrown into the water. The prisoner, who was a man of education and good social position, asked one of these witnesses whether, after six weeks' time, water would remain in the body, and the reply was that there should be some, because 'it can't come out of the body after death but by putrefaction, and there was no putrefaction.' This witness did not appear to have any suspicion that the deceased might have died without swallowing any water, or that the quantity swallowed might have been small, and entirely lost in six weeks' time by transudation through the stomach and lungs. The prisoner was acquitted.

Marks of violence. Marks of violence found on a dead body in the water should always be examined with the utmost care; they may have been produced by the fall into the water, or by the floating of the body up against obstacles after death; or, on the other hand, they may be of such a character that they could not have been so produced, and they may further afford conclusive evidence as to whether they were self-inflicted or not. As an instance of the damage that may be done merely by the contact with the water, mention may be made of the case of a professional diver, who, having twice successfully 'taken a header' from the High Level Bridge at Newcastle, was foolish enough to risk a third attempt. On this occasion, however, he was, as was supposed, caught by a gust of wind in his descent, and fell into the water on his side; when his body was found a few days later it was reported that nearly every bone in it was broken. Fracture of the cervical vertebræ has been sometimes found as the result of a dive into too little depth of water. In one case where this accident occurred the man was taken out of the water insensible, but rallied afterwards, and then said that when he felt his hands touch the bottom he had drawn his head violently back to save it, and had then lost all consciousness. After death there was much ecchymosis about his neck, and the body of the fifth cervical vertebra was found to be fractured. There was no bruise found on the head, and this was held to support the idea that in this case the fracture had taken place from the sudden and violent muscular action; but it should be remembered that people do not feel blows under the water in anything like the same degree that they do out of the water, and that the instantaneous nature of the insensibility prevented any statement on the part of the man himself being reliable.

The medical evidence will often throw light on the question as to whether the drowning was the result of accident, suicide, or homicide. It used to be held that a person could not drown himself in shallow water, that the imperious necessity for air would overcome his resolution; all drowned bodies in shallow pools were consequently looked upon as the handiwork of a murderer, when not obviously the result of accident. This notion has been exploded now. Death in shallow water could only be the result of accident on the hypothesis that the person was intoxicated, or suffering from an apoplectic or epileptic fit, or in a state of syncope. The post-mortem examination would afford in most instances the means for determining three of these, but the question of epilepsy could only be decided by a reference to the history of the person. *[margin: Accident, suicide, or homicide.]*

The finding the body with the hands and feet tied together would exclude the question of accident, as also would the presence of weights tied to the body; but neither of them would furnish any presumption as between suicide and murder, for suicides often take similar precautions to ensure the success of their designs.

CHAPTER VII.

DEATH FROM ASPHYXIA (*continued*).

Death from hanging.

DEATH by *hanging*, as carried out judicially at the present day, is not due to asphyxia—at least in the majority of instances this is not the case, and it is not intended that it should be. A few years ago the leaders of the medical profession in Ireland, headed by the Rev. Dr. Haughton, a member of the profession and an eminent mathematician, thoroughly investigated the subject of the cause of death by hanging, and arrived at certain conclusions with regard to the best method of fracturing the cervical vertebræ, which was admitted to be the most speedy death. They recommended that the short drop then in vogue should be given up, and one of about eight feet adopted; and further, that the knot should be placed under the chin and not at the occiput, as it had previously been. These recommendations were put in force by Marwood when he became public executioner, and have since been followed. Dr. James Barr, who has had a good deal of experience as medical officer to a prison, thus describes the mode of death: The respiratory and vaso-motor centres are at once paralysed; the excito-motor ganglia of the heart keep up its action for some minutes, until finally stopped by asphyxia. The right side of the heart is unable to drive the unoxidised blood through the lungs owing to the cessation

Judicial hanging.

of the respiratory movements, whilst the stagnation of blood in the venous system soon tends to equalise the pressure, and cause the left ventricle to be unable to drive its blood through the systemic capillaries. In this mode of death consciousness is lost almost immediately. The damage done to the spinal column in this form of hanging is always great. Sometimes the transverse ligament is ruptured; this does not often happen; more frequently, but still not often, the odontoid process is broken off. Most frequently, however, the damage is to the third or fourth or fifth vertebræ, and consists in fracture of the body with rupture of the ligaments connecting it with the next vertebra, the spinal cord and vertebral arteries being torn across.

The other causes of death in hanging have been put by Tidy under the following heads: Asphyxia, cerebral hyperæmia, a combination of asphyxia and apoplexy, and syncope. The last is generally due in great measure to fright. Asphyxia is the commonest cause of death in non-judicial hanging, the other two causes being merely instances of inefficient rather than not successful hanging. Death is less likely to result from asphyxia when the cord passes over the larynx than when it lies above it. In this situation the base of the tongue is pushed against the posterior wall of the pharynx, the carotid arteries and jugular veins are compressed, and the pneumogastric nerves; and all probably have some share in the production of death. The subsequent remarks about hanging will not apply to cases of judicial hanging. *Ordinary hanging.*

It is naturally difficult to speak of the symptoms caused by hanging, as it is an experience that but few have lived to narrate, but public performers and others have described the insensibility as coming on very soon, and being preceded by noises in the ears and flashes of light before the eyes. *Symptoms.*

Tidy speaks of three stages: 1. A short stage of semi-consciousness extending from thirty seconds to three minutes. 2. A stage of subjective death but of objective life, varying in time from ten minutes onwards. 3. A stage of general death, lasting until the occurrence of rigor mortis. The expulsion of urine, fæces and semen is found to have taken place in a certain proportion of the cases. The other symptoms may be inferred from the description of the post-mortem appearances.

Treatment. In reference to treatment, the most important thing by far to do is to cut the person down, and loosen the ligature or rope round his neck; if the suspension has not taken place long, not more than a minute or so, this may suffice to save his life. The body should, if possible, be placed in a free current of air, all clothing being removed from about the chest, and artificial respiration should be commenced without delay. Stimulating applications should be made to the extremities, and, where possible, the use of electricity is to be advised, to assist in artificial respiration by faradisation of the phrenics. Warmth should be applied to the limbs, and where there has been great cerebral congestion, bleeding is undoubtedly useful. Complete recovery has been known after at least ten minutes' suspension; and in one case, where half an hour had elapsed, the man was so far restored that he lived for 19 hours. It does not by any means follow, because the immediate treatment is successful, that the person will ultimately recover.

Post-mortem appearances. The mark on the neck. The most important post-mortem appearance in cases of hanging is the mark made on the neck by the constricting body that was used. In the majority of cases this mark is oblique; it is described as a hard, dry, yellowish, horny furrow with superficial ecchymoses in it at times. These characters of the mark naturally depend a good deal upon the kind of ligature that has been used; a small ligature produces a thin deep line, often with extravasation of blood into the subcutaneous tissues. When oblique, the mark is generally lower in front than behind; when there are two marks these usually cross, and are not parallel. If, however, the body was cut down immediately after death, and if a soft ligature had been used, and death had occurred speedily, there might be no mark on the neck at all.

As regards the other appearances after death by hanging, the head will be found to incline away from the side where the knot is; and in cases of suicide, according to Tidy, the commonest position is for the head to be bent forcibly forwards. The face is usually swollen, and protrusion of the tongue is mentioned by some authors. The presence of bloody froth about the mouth and nose has been noticed by some observers; the pupils are generally dilated. Signs of genital excitement are present in a certain proportion of the cases, including thereby the emission of prostatic fluid, the penis being turgid and semi-erect; in women blood is often found about the genital organs.

The internal appearances will be those common to death from asphyxia, and in addition there may be any or all of the following, viz.: fracture or dislocation of the cartilages of the larynx, or fracture of the hyoid bone. There is sometimes deep congestion of the larynx, and the middle and inner coats of the carotid artery have been found divided.

When the body of a person is found suspended the natural question at once arises, Was the death due to hanging? To decide that it was not, is practically to decide that the deceased had been murdered; for a body would not be likely to be suspended by anyone if the death had been due to natural causes, though it is quite likely that a

Diagnosis.

murderer might hope in this way to lead people to suppose that the deceased had committed suicide. It must be admitted that there is no absolutely certain sign, by which it can be at once determined whether the deceased had died from hanging; or whether he had been suffocated first and the body suspended afterwards. The mark on the neck would naturally be closely examined. This is not always ecchymosed, nor is ecchymosis a necessary proof of suspension during life. If the body had been suspended immediately after death, and before there had been time for it to cool down, ecchymosis might be produced. The parchment-like depression may also be produced by suspension after death. The changes that are produced in the subcutaneous tissues by hanging may be produced equally well if the suspension take place after death; so that, if a person had been strangled and the body suspended immediately after death, there might be great difficulty in proving that it was not a case of suicidal hanging. Moral and circumstantial evidence would be required to decide the point. Supposing it to be granted that death had been due to hanging, the next point would be whether this was the result of accident, or whether the case was one of suicide or murder. Accidental hanging is so rare that it may be at once put on one side; it would probably always be perfectly obvious that the hanging was the result of an accident were that the case. As between suicide and homicide, it must be borne in mind that hanging is one of the most difficult, and least likely to be chosen, methods of getting rid of a person, so that to determine that death was due to hanging creates a presumption in favour of the case being one of suicide. It should only be a presumption however, and a medical witness should not allow his judgment to be too much influenced by it. It has been said that in suicide the mark

Accident, suicide or, murder.

on the neck is oblique, while in case of homicide it is circular. No reliance whatever is to be placed on such statements. Additional marks of violence on the body should be examined with the greatest care, for on them the question of murder or not may turn. The marks might be in such a situation, or of such a character, as to render it improbable that the deceased could have inflicted them on himself, or could have had strength enough to hang himself afterwards. A determined suicide, however, often exhibits a greater degree of strength than could have been expected; so that caution must be exercised in saying that any act was an impossibility for any particular person. The discovery of poison in the stomach would rather point to suicide than otherwise, as suicides have often hanged themselves after failing in other ways to destroy their lives. It used to be taught that the discovery of a fracture or dislocation of the cervical vertebræ was a proof of murder; it is now recognised that this depends entirely on the length of the drop, and the weight of the person, and is quite as likely to occur in a case of suicide as in a case of murder. Another fallacy was, that if the hands were tied the case was one of murder; but in some cases suicides have displayed the greatest ingenuity in tying their hands. Again, the position of the body will not afford much assistance, as suicides have managed to hang themselves at a very few inches from the ground. It is worth mentioning that some years ago an American, Dr. Dyer, discovered a transverse fracture of the crystalline lens in each eye of a criminal who had been hung, and in experimenting on dogs, in two cases out of three he found the lens fractured.

Death from Strangulation.—In strangulation, death is due to asphyxia, mechanically produced by constriction or compression of the air-passages from without. It differs from hanging in that death is

<small>Strangulation.</small>

always due to asphyxia. Hanging affords, as has been seen, a strong presumption of suicide; strangulation affords an equally strong one in favour of murder. As a general rule, it will be found that a far greater amount of violence has been used than was necessary merely to destroy life. The symptoms do not call for a separate description. Probably death occurs very rapidly. The treatment would be the same as in hanging, but, unless commenced almost immediately, the chances of success would not be great.

Post-mortem appearances. As regards the post-mortem appearances, the mark on the neck is, of course, the most important. This is generally circular, and goes completely round the neck. It is most commonly below the larynx, but may be anywhere in the neck. Attention should be paid to the existence of lividity or ecchymosis in the course of the mark; very often there will be abrasion of the skin, owing to the unnecessary amount of violence employed. Marks of manual pressure will sometimes be found. It is of the utmost importance, when these are present, to determine which fingers, and of which hand, have left their imprint, as this point may settle the question of suicide or not in a great many cases. The other external appearances do not call for any special remark; they are those of asphyxia, and are rather more marked than in hanging, and, in addition, purpuric spots are often found on the body. As to the internal appearances, the damage done to the deep-seated structures of the neck will be similar to that found in hanging; the mucous membrane of the larynx and trachea will generally be reddened. The lungs may or may not be congested; there may be ruptures of superficial air vesicles, producing superficial emphysema. Some observers give this as a constant sign of strangulation. Sometimes patches of extravasated blood are found throughout the lungs. The condition of the heart varies; the brain may or may not be normal. Involuntary discharge of urine, fæces, or semen is sometimes noted. The blood is dark and fluid. As to whether death was due to strangulation, it may be observed that the fact of finding a ligature tightly tied round a person's neck would certainly create a presumption that there had been a murder, as it is very difficult to suppose that anyone would thus treat a body already dead. The mark on the neck, if deep, with the other evidences of death from asphyxia, such as the prominent eyes, &c., and the purpuric spots, would be almost proof of death from strangulation. The emphysema of the lungs has never been produced on the dead body by strangulation, and may therefore be regarded as conclusive of the strangulation having taken place during life.

Strangulation is occasionally the result of accident. It is not probable that there would be any difficulty in recognising the fact when this was the case. Suicidal strangulation, though not common, is still far from unknown, but, as before stated, it is generally homicidal, and in such case there will usually be marks of a struggle having taken place, unless the murderer took his victim quite unawares. The position of the knot, or the number of times the coil has been passed round the neck, may suffice to determine that the case must have been one of murder. A somewhat celebrated case of imputed homicidal strangulation occurred at Montpellier, in which a man-servant accused his master of having tried to murder him. The man was found lying in the cellar, with a cord coiled round his neck three or four times and not tied at all. There were some marks on his neck, but no ecchymosis and no abrasion of the skin. He was lying upon his left side, and his legs were tied together with a handkerchief

Imputed strangulation.

belonging to his master. According to his own statement, he had been lying on the floor unconscious for eleven hours, having been rendered insensible by a blow on the nape of his neck by his master at that time, and having then been strangled by him. As the condition in which he was found did not coincide at all with the idea of his having been, as he stated, tied up for that length of time, and as there were other signs about him obviously simulated, such as alleged loss of voice, the man declaring that he could not speak above a whisper, the accused was acquitted.

Suffocation.

Causes.

Death from Suffocation.—Death from asphyxia, where the obstruction to respiration is more or less mechanical, but not by violence applied to the throat from without. There are many ways in which death may occur which are included under this head, the majority, perhaps, accidental, some homicidal, and some suicidal. Amongst them may be mentioned the application of the hand or a cloth over the mouth and nostrils, smothering or covering the face with a towel, &c.; the accidental or forcible introduction of foreign bodies into the mouth and throat, the flow of blood into the trachea from the rupture of an aneurysm, &c.; wounds of the throat completely dividing the trachea, the lower end of which gets drawn into the wound so as to interfere with the free entry of air through it; plunging the face and nose into sand, feathers, hay, &c.; œdema of the glottis from boiling liquids, corrosives, &c. Sometimes people have been suffocated in a crowd from pressure on the chest. Sometimes food passes into the trachea, causing suffocation; this is especially liable to occur during vomiting when the person is intoxicated or insensible from an anæsthetic. The bursting of an abscess into the larynx, and the passage of false teeth into the trachea have each been followed by death.

These symptoms and post-mortem appearances differ from those of the other forms of death by asphyxia, rather by negative than positive characters, that is to say, there are no marks on the neck or elsewhere of external violence, and, in the case of very young infants, it is often quite impossible to determine the fact of suffocation with certainty. They are taken to bed by the mother, and in the morning are found to be dead, smothered either accidentally or on purpose 'overlaid,' as it is called. Another way in which it is said that young children are suffocated is that they are given pieces of rag, washleather, or some similar substance to suck, and that this is gradually drawn by the child sucking at it to the back of the mouth, when suffocation is produced; the rag or other substance is then removed from the mouth, and a doctor is sent for, who is told that the child died in a fit.

There is however, one post-mortem appearance upon which a good deal of stress has been laid, as being peculiar to cases of suffocation, and that is the presence of subpleural ecchymoses. They are small, dark, round, well-defined spots, usually seen at the root and base of the lungs. They are not, however, always present in cases of suffocation, and they may be found in death from asphyxia from other causes. Homicidal suffocation is not common. The following case is peculiar in the mode employed. A woman was found to have been suffocated by a cork tightly thrust into her glottis. At the trial it was suggested for the defence that she had been drawing a cork out of a bottle with her teeth, and had thus inhaled it into her larynx. This very ingenious theory was disproved by the fact that the sealing-waxed end was outermost, which it should not have been if the hypothesis of the defence had been true. It appeared that she had been

<small>State of the lungs.</small>

attacked whilst drunk by five men, but there was no evidence to show which of the five was the actual criminal, and so a verdict of 'Not proven' was returned. It was said that the dead body of the man, who was supposed to have been the culprit, was a few days afterwards found in the river.

CHAPTER VIII.

DEATH FROM LIGHTNING.

LIGHTNING kills by producing a profound shock on the brain and nervous system generally. Dr. B. W. Richardson, who has made some experiments on this subject, announced, as the result of his investigations, that the blood is a better conductor of electricity than either water, muscle, or nerve matter, and consequently the course of lightning through the body would be by the blood. Thus, the different ways, in which the effects of lightning are manifested, may be explained by difference in the respective conducting powers of the articles of wearing apparel used by the person at the moment of being struck. A person whose clothes are wet may, if sitting on the ground at the moment of being struck, escape owing to the wet clothes being good conductors, and transmitting the electric fluid direct to the ground. Instances of this have actually occurred.

When a person is struck by lightning, he usually becomes at once unconscious, and death takes place very rapidly (sometimes almost instantaneously), or if not he generally recovers; but sometimes, though not often, death may take place later, as the result of severe burns or other injuries. Sometimes there is no mark at all at the spot where the electric fluid entered the body, and in these cases it has been surmised that it was by the return shock, by which the earth was parting with its excess of positive electricity, that the person was killed.

The clothes are torn sometimes to tatters; watches and other metallic articles are more or less fused; boots, too, are torn into shreds, probably from the presence of nails in the soles. An interesting case was recorded a few years ago in the Clinical Society's Transactions, in which a man who was struck by lightning was found in a field absolutely naked, with the exception of a part of the left arm of his flannel vest. The clothes, most of which were literally in shreds, had been carried a few yards by the wind. The man never lost consciousness at all. When taken to the hospital he was deaf; there were many extensive burns on the abdomen, and down each thigh and leg was a broad streak, terminating at the ankle in an excoriation; it was crimson coloured and indurated throughout its whole length. There was a compound fracture of the right tibia and fibula, about the lower third, and the os calcis was comminuted with a lacerated wound leading down to it.

Amongst the more or less permanent effects, that have been noted in people struck by lightning, but not killed, may be mentioned blindness or impaired vision, deafness, epilepsy, paralysis, and insanity, loss of memory, and headache. The different degree in which people, even when quite close together during a storm, will be affected by lightning, is a matter of general observation, and is well shown in a case narrated by Taylor, where two boys and a man took shelter under a haystack during a thunderstorm. They were all struck by lightning. The man was killed on the spot; one boy had serious nervous symptoms, such as unconsciousness, slow breathing, and a weak and irregular pulse, and recovered consciousness in about five hours; whilst the other boy was so little injured that twenty minutes afterwards he was able to walk.

It has been said that rigor mortis does not occur after death by lightning stroke. This is an error. In the cases where it was supposed not to have supervened, it is probable that it came on, as it often does, very soon, and passed off early. It does not always come on so early, however, as in the case just alluded to rigor mortis did not appear until 14 hours after death. Another fallacy, which experience has corrected, is the notion that the blood does not coagulate in these cases, though no doubt its coagulation is sometimes delayed. The external appearances have incidentally already been described, and, as regards the internal, the most prominent are congestion of the various viscera, and extravasation of blood beneath the skull.

CHAPTER IX.

DEFINITION AND CLASSIFICATION OF WOUNDS.

Definition of a wound. THE use of the word 'wound' seems at first sight to imply the infliction of something more than a mere blow; it seems to necessitate that there should at least be some abrasion of the skin; but such is not the case. In the eye of the law, no such distinctions are recognised, and the statute is so worded that no external damage is necessary; nor is even the use of a weapon essential, as would very naturally be thought, for it seems somewhat anomalous to speak of wounding a person with the fist. But the comprehensiveness of the statute does away with any difficulties of this sort, for it runs: 'Whosoever shall, by any means whatsoever, wound or cause any grievous bodily harm to a person,' &c.; so, then, neither the use of a weapon, nor the presence of external damage, is necessary to a charge of breaking the law on this head.

There are several points to be considered under the head of Wounds. In the presence of a dead body, the question that first arises is, whether any wounds, that may be found, were caused during life or after death; and, if the former, then the further question arises as to whether they were the cause of death, and whether they were self-inflicted, resulted from accident, or were inflicted by someone else. In the case of a living person who may be found wounded, the probable danger to life of such wound or wounds will have to be estimated; and here possible contingencies, such as tetanus, pyæmia, &c., are not to be taken into consideration, but only the likelihood there is that the wound in itself will prove fatal.

Various classifications of wounds have been proposed. Some writers, such as Beck, divide them into slight, dangerous, and mortal. There are, however, many objections to such a plan, though at first sight it might seem to have much to recommend it. There are so many considerations to be thought of, in regard to individual peculiarities and susceptibility, that this division would be found inconvenient in practice. It is, of course, true in a general way that wounds of the medulla oblongata, of the pneumogastric nerves, or of the heart and large blood-vessels, are likely to prove speedily fatal; and that wounds of the lungs, of the liver, spleen, diaphragm, peritonæum, and brain may fairly be classed as mortal.

Classification.

Others have divided wounds into incised, punctured, lacerated, and contused—a classification which possesses this advantage, at any rate, that it facilitates the description of the wound. An incised wound is what, speaking ordinarily, would be called a cut—that is, it is a wound, in which there is a solution of continuity, without any loss of substance. A punctured wound suggests the idea of a wound caused by some sharp-pointed instrument, which has penetrated more or less deeply, but has not made a large wound of entrance. A lacerated wound is an incised wound the edges of which are irregular, whether from loss of substance or other causes; and a contused wound implies considerable bruising of the parts around, with extravasation of blood into the tissues.

To distinguish between a wound inflicted after

Ante and post mortem wounds. death and one made during life, certain general principles should be borne in mind, which will go far towards solving the question in any given case. If there are any signs of repair about the wound, such as the formation of pus, or if there are signs of gangrene (though it is important not to confound these with commencing putrefaction), the wound must have been inflicted at least twelve hours before death, or longer if the person was in a very feeble state of health. On the other hand, a wound inflicted more than twelve hours after death could never be mistaken for one produced during life; there would be no blood and no clots in the neighbourhood of the wound, and the edges would not be retracted. If the wound took place within twelve hours before death, the arteries in the surrounding parts would be empty; there would be clotted blood in the margins of the wound, which would be retracted, and the subcutaneous tissues would be infiltrated with blood. In a wound made immediately after death, all the above characters might be present, but some good reason should be forthcoming why wounds should be inflicted on a body so soon after death, if it was really true that they had been so inflicted. The longer the time that had elapsed since death, before the infliction of the wounds, the less difficulty there would be in deciding.

Dr. Taylor and Mr. Key made some investigations on this subject, of which the former gives the following account:—Two minutes after a limb had been amputated, a deeply-incised wound was made in the calf. The fact, that in an amputated limb the blood would have drained away, and therefore the conditions not be precisely the same as those of a wound inflicted on an unsevered limb after death, was duly noted. At the moment that the wound was made, the skin contracted considerably, causing a protrusion of the fat beneath; the quantity of blood which escaped was small, and the cellular membrane, by its sudden protrusion forward, seemed to prevent its exit. The wound was examined after the lapse of twenty-four hours; the edges were red, blood-stained, and everted; the skin was not in the least degree swollen, but merely somewhat flaccid. On separating the edges, a small quantity of fluid blood escaped, but no coagula were seen adhering to the muscles. At the bottom of the wound, however, there was a small quantity of coagulated blood; but the coagula were so loose as to readily break down under the finger.

In a second experiment, ten minutes after the separation of the limb from the body, a wound of similar extent was made on the outer side, penetrating to the deep-seated layer of muscles. In this case the skin appears to have lost its elasticity, for the edges of the wound became but slightly everted, and scarcely any blood escaped from it. On examining the leg, twenty-four hours afterwards, the edges of the incision were pale and perfectly collapsed, presenting none of the characters of a wound inflicted during life. At the bottom of the wound, and enclosed by the divided muscular fibres, were some coagula of blood, but evidently fewer than in the former experiment. Other experiments were performed at a still later period after the removal of the limb, and it was found that, in proportion to the length of time suffered to elapse before the production of a wound, so were the appearances less distinctly marked—that is to say, less likely to be confounded with similar injuries inflicted upon a living body. When the incised wound was not made until two or three hours after the removal of the limb, although a small quantity of blood escaped, no coagula were found.

It is worth mentioning that hæmorrhage is more

certain to be produced by an incised wound that involves a large vessel, than by a lacerated or contused one, and it is not an uncommon thing for a limb even to be torn off, *e.g.* in a machinery accident, without any very great loss of blood.

<small>Ecchymoses not always due to violence.</small>

The mere fact of an extravasation of blood under the skin—an ecchymosis, as it is called—must not be taken as a proof of violence, for in certain diseases, as in purpura hæmorrhagica, for instance, it is quite common for large extravasations of blood to appear, without any history of accident, which in a few days cannot be distinguished from ordinary bruises. Again, a blow inflicted during life may give rise to an extravasation of blood, of which there may be no signs even at the time of the death of the person, and a few hours afterwards these may be well marked, for it is to be borne in mind that there is nothing to prevent extravasated blood from diffusing itself through the tissues after death, or to prevent its undergoing those gradual changes after death which during life give rise to the familiar appearances at different stages of a bruise. The late Sir Robert Christison, who paid a good deal of attention to the subject of contusions in the dead, said that blows on the skin within two hours of death produced appearances, which could not be distinguished from those by a blow inflicted during life. It should be remembered that blows inflicted during life do not always produce extravasation of blood, and that the blood when so extravasated does not always coagulate, so that neither the absence of an extravasation of blood, nor the presence of fluid blood about a contused wound, can be accepted as conclusive evidence of the wound having been inflicted after death.

Woodman and Tidy have put the differential characters of wounds inflicted before or after death in the following tabular form:—

Incised Wounds.

In the living.

1. Edges sharply cut and everted, the skin and muscles being retracted.
2. Bleeding copious, and generally arterial.
3. There are clots.
4. There is a good deal of staining or diffusion of blood in the muscular and connective tissues.
5. After some hours or days there will be signs of repair or inflammation.

In the dead.

1. Edges close, and not everted.
2. Bleeding absent or scanty.
3. There are no clots in most cases; sometimes a few small clots.
4. There is very little or no staining or diffusion of blood in the tissues of the wound.
5. There will be no attempt at repair, and no signs of inflammation. There may be signs of putrefaction.

Contused Wounds.

In the living.

1. There is swelling, and after a few hours, or a few days if deep-seated, the skin changes colour, particularly at the edges.
2. There is effusion of liquid blood and lymph in the deeper parts, and coagula form.
3. The swelling subsides and the colours fade after some days, or, in some cases, weeks.
4. Abscesses may form, or ulceration, sloughing, or erysipelas set in.

In the dead.

1. There is very little swelling or change of colour.
2. Very little blood is effused. There are hardly any clots.
3. There are no rainbow-like or prismatic changes of colour.
4. No abscesses form, and no erysipelatous or gangrenous changes are met with.

Lacerated Wounds.

In the living.

1. There will be more hæmorrhage and staining from the blood at first.
2. After a few hours or days there will be suppuration or other sign of repair; inflammation or gangrene may also supervene, as in incised wounds.

In the dead.

1. There is hardly any hæmorrhage or staining, unless large veins are torn across.
2. No evidences of repair or inflammation or gangrene can be detected.

Punctured Wounds.

In the living.	In the dead.
1. There is generally considerable bleeding during life.	1. There is little or no hæmorrhage or coagulation of blood.
2. Repair or inflammation or necrotic changes set in after a few hours or days.	2. There are no evidences, either in or around the wound, of repair, inflammation, or death of the tissues.

CHAPTER X.

CAUSATION OF WOUNDS, BLOOD STAINS, ETC.

THE examination of the clothes of a wounded person should never be omitted. Often the evidence afforded by them is of the most vital kind. Thus in one instance a man accused some others of attacking him, and showed a cut on his forehead in support of the charge. He said he was wearing his cap at the time, and pointed to a cut upon it as additional evidence, but the cut on the forehead and on the cap sloped in opposite directions, therefore the prosecutor's statements could not all be true, and it eventually turned out that the wound was self-inflicted. In another instance an old woman was found dead in her bed, and a fellow-lodger was accused of having caused her death, her skull having been found to be fractured. There was a history of her having been seen to fall in the road the night before, but not much attention was paid to this until her bonnet was produced with indentations exactly corresponding to the fracture of the skull, dust being ingrained in the depressions. The history of the fall was therefore substantiated in the strongest manner possible. *Evidence from the clothing.*

It having been decided, then, that a given wound was inflicted before death, and was the cause of death, it becomes necessary to determine whether it was the result of accident, suicide, or homicide. *Suicide or murder.*

It is not wise on this matter to lay down any very hard-and-fast laws, as there are but few forms of injury, of which it would be possible to say that they could not have been self-inflicted, or the result of homicidal violence, as the case might be. The points, upon which the question would have to be determined, would be the situation of the wound, its nature and its direction. Suicidal wounds are generally on the front part of the body, that is, in parts most accessible to the person, for the suicide would naturally consider that he would have a better chance of effecting his object, when wounding himself in front, than by giving himself what would be rather a chance wound behind; and, besides, the heart, the mouth and temples, which are the favourite regions for suicides, are all more or less towards the front of the body. Suicides often shoot themselves in the mouth; in such a case the muzzle of the barrel would be placed inside the mouth, so that, if there was much damage done to the teeth by the bullet, there would be a presumption that the wound might not have been suicidal. It is hardly possible to suppose that a person could be accidentally shot through the mouth without a good deal of damage being done to the jaw. Should the fingers be found blackened with powder, as in the case of a muzzle-loader, or the hand grasping the pistol, razor, or other weapon, the fact would furnish, as has been already said, almost conclusive evidence of suicide. Suicides, especially when insane, sometimes inflict the most unlikely wounds upon themselves. Thus, one man tried to kill himself by running head foremost against a wall several times. Another was seen in the act of striking himself on the head with a cleaver.

Evidence from the characters of the wound. A suicidal wound in the throat will be from left to right, or *vice versâ*, according as the person was right- or left-handed. A clean cut is very generally accepted as presumptive evidence of suicide, and rightly so; but nevertheless a suicidal wound may be irregular, either from indecision, or weakness, owing to the amount of blood already lost. A clean-cut wound in the throat could hardly be the result of homicide, unless the person was under the influence of some powerful drug, or very soundly asleep at the time of the assault. From the nature of the wound alone it is often possible to form a correct opinion as to the probability of accident having been the cause or not. Less often it is possible to decide between murder and suicide on this ground alone; but in one instance, where a farmer was found with his throat cut from ear to ear in a semicircular manner, just as the throats of animals are cut by butchers, the murder was traced to a man who had been a butcher. The direction of the wound should always be closely studied, as it may throw much light on the causation of it. Thus suicidal wounds in the throat generally terminate in the skin; that is to say, the skin is the farthest point wounded, the weapon not being carried beneath it.

A very recent instance has been afforded of the necessity of paying close attention to the direction of wounds in the Uxbridge murder. The deceased, who had been married for many years and lived alone with his wife, was found dead in his bedroom one evening with four bullet wounds in different parts of his body. His wife made statements, as to what had happened, to several people, which did not agree very well with each other. There was no reason why the deceased should have committed suicide, nor was there much reason why his wife should have wished to murder him. As there was no one else in the house, it was clear that the case was one of murder or suicide, or that the deceased had been shot during a scuffle. The medical evidence was

The direction of the wound.

practically conclusive that the case was not one of suicide, and the wife was convicted of wilful murder. Three of the wounds were in the front of the body, and though it was highly improbable that they had all been self-inflicted, it was not impossible. One of them was in the right cheek, causing most serious injuries to the superior maxilla, whilst another was on the left side of the chest, the bullet having passed through the heart and lodged in the liver. None of the wounds in the front of the body showed any trace of having been fired from quite close. The wound in the back was close to the left shoulder, and the clothes around showed that the muzzle of the pistol must almost, if not quite, have been touching them; the ball had passed through the chest nearly in a straight line from behind forwards, and it was the direction of this wound that was inconsistent with any theory of suicide or accident, as it was proved by the medical witnesses, that it could not have been inflicted whilst the pistol remained in either hand of the deceased.

The best illustration, that can be given of the advantage of studying the direction of the wound carefully, is the case where Sir Astley Cooper was called in to see a gentleman, who had been shot, whilst sitting in his chair in his own room, by an unknown assailant. Sir Astley, after careful investigation, came to the conclusion that the pistol must have been fired by a left-handed man. This led to suspicion of the only left-handed person about the house, who subsequently confessed to the deed.

The fact of a very extensive wound in the neck should not lead to a hasty presumption of murder, for determined suicides sometimes inflict a remarkable amount of damage upon themselves; thus a person has been known to divide all the structures in the neck and even notch the vertebræ.

In all cases it is very desirable that there should be no disturbance of the body, or of any of the articles of furniture in the room, before the arrival of the medical man. Attention should always be paid to the position of the weapon, as suicides often manage to throw it away after inflicting upon themselves a fatal wound, and the question will arise whether, after such a wound had been given, the person could have preserved sufficient strength to have thrown the weapon to the spot where it was afterwards found.

Stabs and punctured wounds may arise accidentally from some sharp-pointed instrument carried in the pocket. In such cases the direction of the wound is from below upwards, and such a direction always suggests an accident, though it is not by any means conclusive. In the subjoined case the decision as between accident and homicide turned entirely upon the direction of the wound. A woman was charged with murder under the following circumstances:—A fall was heard in the dining-room, and the deceased was found on the floor dying from a wound in his chest. A large table-knife was found near the deceased, with which the wound had evidently been caused. There was evidence of a quarrel, but the question was, whether the wound could have been received accidentally during a scuffle, or whether it was obviously not so caused. The wound penetrated the left side of the chest, between the sixth and seventh ribs, reaching and wounding the apex of the heart. The wound was downwards, forwards and inwards, in one uniform direction, straight from end to end, and had never changed its course. There was not the slightest upward tendency in the wound. The surgeon who gave evidence at the trial said that the wound being from above downwards the pressure must have come from above, whereas in the case of a falling

I

body the pressure would come from below. The wound must have been inflicted by another with considerable force, and he could not frame any theory by which it would have been the result of accident.

The presence of several wounds on a body is not necessarily proof of murder, though the presence of many dangerous ones would practically be so; for instance, of several deep stabs in the chest; but a suicide may inflict many slight wounds on himself, before the fatal one, through lack of resolution. Also the presence of wounds inflicted by different weapons is not evidence of homicide. Suicides often try various methods in succession. In the case of several wounds, it would be proper to endeavour to determine the order in which they were received, and which was the fatal one, though it is possible that more than one may have contributed to the fatal result.

The importance of noticing the position of the body, and of the surrounding articles of furniture, and especially of the weapon, if one be present, has already been alluded to. In the case of a wound that from its nature must have proved rapidly fatal, if the weapon, with which such a wound might have been inflicted, cannot be discovered, this would be evidence against the theory of suicide, but it must be remembered that people sometimes survive, even wounds of the heart, for a considerable time, and retain power enough to perform many voluntary acts.

The subject of blood stains will be dealt with separately, but it may here be observed that from the presence of blood, on the furniture or about the room, valuable evidence may often be derived pointing to the probability of murder or suicide, as the case may be. If the person be found lying in a pool of blood, and there be no marks of blood elsewhere in the room, the inference will be that the wound was given on that very spot, and that death was tolerably rapid; and if under such circumstances, the weapon, with which such a wound might have been given, cannot be found, there will be a strong presumption that the case is one of murder. The weapon, if found, must be examined for blood stains with the most scrupulous care.

Besides blood, other substances may be found adhering to weapons, which may be of great use as evidence. Such are hairs and fibres. If hair is found, it may be possible to prove that it is the hair of the deceased person, which, in the case of a wound about the head, would be the best possible evidence that such a wound had been inflicted with that very weapon. In other cases fibres, which have been found adhering to a knife, have been proved to belong to some article of clothing worn by the deceased at the time of the murderous assault. The microscopical characters of hair will be found in the text-books on histology. It must be remembered that the hair of animals must not be confused with human hair. As a rule, the former is coarser, shorter, thicker, and less transparent. The fibres that might have to be distinguished are those of cotton, linen, silk, or woollen fabrics. Briefly their characters are as follows:—*Cotton* appears as a flattened band of more or less spiral form. *Linen* is of a rectilinear form, with jointed markings at unequal distances, the fibre tapering to a point. *Silk* has a regular cylindrical form, and there are no markings on the surface; it has a strong refracting power on light. *Woollen* fibres are irregular, contorted, of unequal thickness, and they have peculiar markings, of an imbricated character, on the surface.

In the case of firearms, when the bullet has not been extracted, supposing the wounded person not

Examination of hair or fibres on the weapon.

to die, great caution should be taken not to express an opinion that a pistol, found on the suspected person, was the one with which the bullet was fired. An instance of the necessity for such caution occurred in 1859. A person was shot in the left temple. The ball could not be extracted at the time, and he did not die at once. A man was arrested on suspicion, and two pistols of different sizes were found on his premises. No comparison could be made, and so the prisoner was remanded. Nine months later the wounded man died, and it was then found that the bullet would not fit either pistol.

Evidence from the presence of blood stains. The presence of blood stains on the assailant is naturally regarded as very good evidence, but the absence of such stains should not be accepted as conclusive of a person's innocence, for the murder may have been committed from behind, in the case of the throat being cut. A case of this kind occurred in the Eltham murder, as it was called, when a young woman was found dead with severe injuries to her head, inflicted by a plasterer's hammer, the temporal artery amongst others being divided. A man was arrested on suspicion, against whom there was a good deal of evidence, but the judge who tried the case summed up strongly in his favour, on the ground that the small quantity of blood on his clothes was inconsistent with the idea of his having committed the murder. Another point, too, to be taken into consideration, before letting a prisoner off on the plea that the clothes did not show enough signs of blood, would be to make quite sure that the clothes, worn by the prisoner at the time of the supposed commission of the crime, had been examined. It was the neglect of this precaution in the case of the Road murder, that caused it to remain so long enshrouded in mystery. The culprit, a girl aged 16, had murdered her infant half-brother in the middle of the night, and, from the nature of the wounds, it was not possible that her nightdress could have escaped being stained with blood. It appeared at the trial, which took place soon after, that she had three nightdresses, but that one had been lost at the wash, as she alleged, during the very week after the murder. This was denied by the laundress and her daughter. The girl had in reality hidden the blood-stained one on the night of the murder, and put out another for the wash, which she had taken back after the things had been counted, and had burnt the blood-stained one a few days later. In this way there seemed to be some truth in her assertion, that the missing one had been lost at the wash.

Blood stains may be recognised by means of the microscope, the spectroscope, or by chemical tests. There are several substances, which give stains having more or less resemblance to those of blood: such are red dyes; these, however, being fixed by mordants, are insoluble in water. Iron-moulds are often mistaken, but they, too, are insoluble in water. If glacial acetic acid be added to the stain of an iron-mould, and shortly after some tannic acid, the inky colour of the stain will reveal the presence of the iron. The juices of red fruits produce red stains, which, moreover, are soluble in water; these are turned bluish-green by ammonia, and logwood and other such dyes are turned crimson by ammonia. *[Stains resembling blood.]*

A fresh blood stain is of course red, but in a few days it fades to a dull brick-red colour, and any opinion as to its age after that period is rather a matter of guesswork. In criminal cases means will often have been taken to remove the stains, but, except by washing in cold water, this is not very likely to have been accomplished; and washing in warm water, the use of alum, &c., only serve *[Characters of blood stains.]*

to make the stain more permanent. When a blood stain on clothing is examined under the microscope, it will be seen that each fibre is surrounded by a portion of blood-clot, and has a shining glossy appearance. If the suspected spot be cut out and placed in water, the solution thus obtained will give the tests for blood. An old blood stain on a knife, or other weapon, will very closely resemble rust; it should be scraped off and put into water, when, if any blood be present, the solution will have a reddish tinge. If the solution is yellowish, and gives a blue colour with ferrocyanide of potassium, the stain was due to the juice of an orange or lemon.

It is not wise to attempt any distinction between arterial and venous blood; or between that of a man, woman, and child; or between menstrual and ordinary blood. The chemical examination of a blood stain should be conducted as follows: If on linen or other article of clothing, the stain should be cut out and suspended in a tube containing distilled water; if on some piece of furniture, it should be carefully scraped off and similarly treated. The solution thus obtained will be fairly clear and have a reddish colour; on boiling, the solution will become more or less turbid, according as the quantity of albumen present is large or small; if large, the whole mass might coagulate; if small, there would only be a slight turbidity; the red colour disappearing and the whole becoming of a dirty green colour, the uppermost portion, perhaps, being somewhat yellow. If some liquor potassæ be now added the solution becomes clear again, but is brownish-red by refracted light, and green by reflected light. Solutions of blood do not become crimson, or undergo any change of colour, on the addition of ammonia: this test serves to distinguish blood from almost any other red solution. The most delicate chemical test, how-

Chemical examination for blood.

ever, is that with tincture of guaiacum and ozonic ether. It is best performed by adding to a drop of the tincture of guaiacum a few drops of ozonic ether until a clear solution is obtained; a few drops of the solution containing blood should then be very carefully poured into this, when a blue colour will be produced, owing to the action of the colouring matter of the blood on the resin, though it will not produce this except under the influence of the peroxide of hydrogen as contained in the ozonic ether. This test is one of extreme delicacy, and will give a good reaction with a solution of blood, which to the naked eye is perfectly colourless.

Spectrum analysis affords a very valuable means of detecting blood. The spectrum of fresh blood (oxyhæmoglobin) shows two bands at the yellow end of the green portion of the spectrum; if the blood be not so fresh but somewhat deoxidised, the two bands will have disappeared, and be represented by a single broader band between the two; this is the spectrum of venous blood, and if this be shaken up with air, the two original absorption bands can be made to reappear. Deoxidised hæmatin gives two fresh bands in the green portion, nearer the violet end of the spectrum than those of hæmoglobin, that nearest the violet end is nearly double the width of the other. *Spectrum analysis.*

The microscope, after all, affords the most trustworthy means of detecting blood, and no examination would be complete, or satisfactory, in which its use had been neglected. In the case of fresh blood there is not the slightest difficulty in recognising the coloured corpuscles of the blood, but with an old blood stain it is by no means so easy. Corpuscles seen under these circumstances are more or less shrunken, and have lost their shape; the best method to detect them is, if the spot be very small, to breathe on a slide and then place the scrapings *Microscopical examination.*

on this, and breathe upon them before putting the cover-glass on; this should be done quite lightly, it should not be pressed down, for fear of squeezing the corpuscles; or, if the spot be larger, it may be macerated in a very small quantity of water, the object of using a very small quantity of water being to avoid the danger of causing the corpuscle to swell up too much and burst, when all chance of recognising it would be at an end. It has been recommended to macerate the suspected stain for half an hour in a solution of sulphate of soda, rendered alkaline by the addition of caustic soda. Hæmatin crystals may also be seen under the microscope; they are in the form of rhomboidal plates or columns, and are brownish-red in colour. They are soluble in potash, and slightly so in ammonia and dilute sulphuric and nitric acids, but insoluble in the other acids, and in chloroform, ether, alcohol, and water. They are, however, difficult to obtain, and, as a matter of fact, very rarely seen.

But having thus proved that certain stains are undoubtedly due to blood, it will at once be asked: Is it certain that these are the stains of human blood? Several years ago some Frenchmen endeavoured to prove that by mixing blood with sulphuric acid, the smell of the animal to which it had belonged could be detected, and they professed to differentiate between the blood of man, and of the various animals in this way. Although it may to some extent be true that blood does evolve a peculiar odour in the presence of sulphuric acid, yet at most this could only be applied to fresh blood in considerable quantity, and not to a small and stale stain, so that the test may be absolutely disregarded. The microscope alone can distinguish between the blood of man, and that of animals; for details on this point the student is referred to works on physiology, and comparative anatomy. It may be sufficient here to remark that, as regards shape, in all mammals, except the camel, the coloured corpuscles are round, and that in birds, fishes, reptiles, and camels the corpuscles are oval; and, further, that the coloured corpuscles of all mammals have no nucleus, whilst those of birds, fishes, and reptiles, have a central nucleus. As regards size, there is no relation between the size of the animal and the size of the corpuscle; nor in the case of the same animal is there any difference at different ages. The coloured corpuscles of man are larger than those of the domestic animals.

_{Characters of the blood corpuscles.}

CHAPTER XI.

THE CAUSE OF DEATH FROM WOUNDS.

IN all cases of death from violence it is of importance to consider what was the cause of death, and, in the presence of several wounds, to decide, when possible, to which the fatal result is to be attributed. The wounds may have been inflicted by different persons at different times; it is therefore obvious, in the interests of justice, that a certain knowledge of the cause of death should, if possible, be obtained in every case. If there is any doubt on the matter, then it is only fair to the defendant that the doubt should be stated at once. Wounds may be said to prove fatal either directly or indirectly. In the former class of wounds there are three ways in which death may occur.

Wounds directly fatal.

1. *Hæmorrhage.* — When a large vessel is wounded, a person may bleed to death either externally, or internally into the peritonæum or pleura; the death in either case would be due to syncope. If the bleeding took place into the cavity of the cranium, as in wounds of the middle meningeal artery, death would be due to compression of the brain rather than to actual loss of blood, and the same applies when bleeding takes place into the pericardium, the heart's action being impeded and death due to that, and not to syncope. Much depends, too, upon the rapidity with which the blood

is poured out; thus, a person may lose by gradual oozing in the course of a few days an amount of blood which, if poured out all at once, must have been followed by immediate death. Children are more susceptible to the loss of blood than adults, and those who suffer from hæmophilia or the hæmorrhagic diathesis are liable to bleed to death after the most trifling injury. After death from loss of blood, the skin and internal organs will be found very white and waxy-looking; the vessels, in the immediate neighbourhood of the wound, will be open and empty, and putrefaction is delayed.

2. *Shock*.—This would include cases where there was no evidence of the blow or other injury, as in the case of sudden death from a blow in the epigastrium, of which many instances have been recorded, and the only explanation of which that has been offered is that some impression has been produced upon the solar plexus or cardiac ganglia. Cases of concussion of the brain proving fatal would come under this head, and also railway accidents where no signs of mechanical injury can be discovered; deaths from lightning or electric shock might be classed in this group. Burns and scalds, too, prove fatal at the time or soon after through shock. A person may have received a great many wounds, none of them dangerous in themselves, but which, when coming all together, may cause death from shock; such a case was that of the schoolboy who was flogged to death by his master.

3. *Mechanical Injury*.—Death from this is probably the result partly of loss of blood and partly of shock; to this class belong severe wounds of the head, thorax, or abdomen, such as are met with in railway accidents, or in persons run over in the streets.

Sometimes a wound proves fatal for want of good medical advice. This would not be allowed any weight in the defence in a trial for murder, even though it could be shown that the wound would, in all probability, not have proved fatal under better medical treatment. There are other things that must also be borne in mind. A wounded person may die soon after receiving the wound, but from natural causes, and his death may in no way have been influenced by the wound. Supposing a person to have been the subject of some infirmity at the time of receiving the wound, it would become a question for the medical men to decide as to whether the death, which may be admitted to have been inevitable, was in any way accelerated by the wound, for it is clear that to accelerate a person's death is to cause it. The law does not require the manner, or the means by which the person's death was caused, to be set forth in the indictment. It must not be forgotten that death may follow a slight injury—a kick on the shin might be followed by acute necrosis and pyæmia. Again, there might have been some latent disease, which a slight blow would kindle into a dangerous and even fatal condition; thus, a boy was hit upon the head by a schoolmaster, and died a few days later of meningitis, when it was discovered that a cyst in the occipital lobe, which had never given rise to any symptoms, had been ruptured by the blow and had set up a fatal meningitis.

Death at a remote period. A wound may prove directly and necessarily fatal from its nature, but not immediately so; and the wounded person may linger for some months, and yet his death be caused as truly by the wound as if he had died immediately on receiving it. Unless, however, he die within a year and a day of receiving such a wound, the law will not allow anyone to be tried for the crime of homicide. From a medical standpoint this seems decidedly unjust, though of course, as a rule, wounds that are likely to prove fatal in themselves generally do so within a few weeks or months of their infliction. The following is an instance of the unsatisfactory nature of such a time limit:—A man came under treatment for vomiting, difficulty of breathing, and great pain in the left chest. After his death it was found that the colon and omentum had passed into the thorax through a hole in the diaphragm, and the intestine was adherent to a scar in the diaphragm, an evidence that the lesion was of some standing. It was afterwards ascertained that, during a quarrel fifteen months previously, the patient had been stabbed in the lower part of the left chest, and that he had since had two attacks of intestinal obstruction, but on each occasion had apparently recovered.

Of wounds which prove fatal *indirectly* it may be said, that the cause was either unavoidable or avoidable.

Wounds indirectly fatal.

It has already been noticed that the plea, that with more skilful medical advice the person would not have died, is not accepted by the law in mitigation of the crime of murder, and rightly so; but it must be remembered that the prisoner's counsel may impute the death to the treatment, and not to the original injury, so that a practitioner should be careful to follow the recognised lines of treatment, and give his reasons for the various steps of that treatment.

Again, it may be that the wound has proved fatal owing to the ignorance or obstinate stupidity of the injured person, or both. He may have neglected to take the slightest precautions, which ordinary prudence would require that a person in his state should take, and he may have resolutely refused to submit to treatment, which he was warned would be the only means of saving his life, but in any such case, if the original wound was given with malicious intent, the crime is all the same, whether the person take care of himself or not. Or, again, it often happens that, owing to some diseased state or congenital abnormality, a wound which, in an otherwise healthy person, would not have been productive of much harm, leads to a fatal result. In this case, too, the motive with which the blow was given would be allowed to be taken into consideration, though the law is framed upon the principle that no one is bound to keep his body in such a state as to warrant an assault upon him.

Indirect causes of death. The indirect causes of death are tetanus, pyæmia, erysipelas, gangrene, and secondary hæmorrhage. Their appearance must be clearly traced to the injury, and the greater the interval that has elapsed, since the infliction of the wound and their onset, the greater will be the difficulty of proving this point. In regard to tetanus, the possibility must be borne in mind of its being due to some other wound, or of its being idiopathic, by which is meant that the cause cannot be made out, not that there is no cause—no case of tetanus can possibly occur without adequate cause. A very good instance of the necessity of minutely investigating the cause of a case of tetanus occurred a few years ago. A boy who had been quarrelling with another boy overnight, and had received a kick in the back, was taken to the hospital next day, with a history that about two hours after the blow he had noticed stiffness in the jaw, and had passed a restless night complaining of pain in the jaw, the stiffness having gradually increased. The other symptoms of tetanus supervened, and he died on the fourth day after the blow. On careful inquiry, however, it turned out that six days before the blow he had run a rusty nail into his foot, and the wound thereby caused had only closed on the day on which he had the encounter in the street, and on examination of the foot a small scar was found on the ball of the great toe. This wound was in all probability the cause of the tetanus, the blow having had nothing whatever to do with it; but, had it not been for the very careful investigation made into the whole case, it is possible that the boy, who struck the blow, would have been put on his trial for manslaughter.

Just as much care will be requisite in the case of erysipelas, which would probably never supervene if the wound could be protected from all external influences. In the case of erysipelas of the face, it must be remembered that there is a so-called idiopathic erysipelas of the face, by no means an uncommon disease, so that this should be excluded before a wound is incriminated as the cause.

The death of the late President Garfield affords a good illustration of death at a late period from a wound. He had been shot in the right side, the bullet impinging upon and fracturing the eleventh rib. The track of the bullet could never exactly be made out during life. Some months after receiving the injury the patient died, and it was then found that the bullet had passed through the body of the first lumbar vertebra, and had become lodged behind the pancreas. Death was due to extravasation of blood into the peritonæum from a rupture in the splenic artery, and it was evident that the artery must have been damaged by the bullet, and that changes had ensued in its coats at the point where ultimately the rupture occurred.

It may happen that the person does not die from the wound, but from the treatment intended to cure the wound. It would be for the operator to show, that the wound would have been likely to prove fatal without an operation, and that the operation he performed was the right one under

Death caused by the treatment.

the circumstances, and performed with due skill. A mistake in diagnosis is always possible; for instance, supposing an aneurysm was mistaken for an abscess and opened, the result, in all probability, would be the almost instant death of the patient. In such a case, if the surgeon was able to show that he 'had exercised a reasonable amount of care, in endeavouring to make his diagnosis, and that the treatment he had adopted, on the strength of that diagnosis, was correct and recommended by the best authorities, the man who gave the wound would be responsible for the death.

It occasionally happens that the friends of a patient or the patient himself, dissatisfied with the results of treatment, bring a charge of malapraxis against the practitioner who has been in attendance. Such charges are especially apt to be brought in cases of fractured limb, where the result is not as good as might have been wished for; very often this indifferent result is entirely due to circumstances beyond the control of the doctor, such as the nature of the accident, or the supervention of delirium, or some other complication on the part of the patient, which has hindered the efficient and thorough carrying out of the treatment. In such a case of course the charge could not be sustained; but if it can be shown that splints or bandages have been carelessly applied, or left on too long, or that the treatment has not been conducted with skill in some other respect, the jury will probably award substantial damages to the plaintiff. The law on this subject was well expounded by a judge, who told the jury that 'they had a right to expect from him (the defendant) the usual and ordinary amount of care, skill, and attention, which it was reasonable to suppose he would possess; and if, in the discharge of his duty, he applied his professional skill and knowledge to the best of his ability, then,

THE CAUSE OF DEATH FROM WOUNDS.

however unfortunate the termination of the case, he was not to be visited with an action to mulct him for damages. Such a step would be most unjust, and have an injurious tendency, and would check that independence of action so necessary for medical men to possess.'

[A case of considerable importance has recently occurred in this country, where a father brought an action for damages against two medical men, in that they had advised him to suck the tracheotomy tube of his child, who had just been operated upon for diphtheria. He became infected with the disease himself, and his contention was, that he had not been duly warned of the risk he was running. The case was summarily dismissed in more than one court. In America, recently, a curious charge of malapraxis was brought against a practitioner by a lady, for taking a young man into her room, at the time of her confinement, who was neither a medical man, nor a student of medicine, and a verdict with damages was given in her favour.]

There is one other point, in connection with the subject of death from the operation, and not from the injury, which must not be overlooked. Now that chloroform and other anæsthetics are so freely administered, even in trifling cases, it would seem to be certain that, sooner or later, a person will die during the administration of chloroform for some such accident as a dislocation, or something of that kind, which could not be said to be in the least degree likely to prove fatal. The question will then arise—it does not appear ever to have been raised hitherto—as to the share, in the causation of death, of the person who gave the blow, supposing the injury to have arisen in this way. As against him it will be urged with perfect truth that, had it not been for his act, the person would not have died, as there would have been no necessity for the

Death from the anæsthetic, and not the injury.

administration of the chloroform; and, on the other hand, he will reply that, had the chloroform not been administered, his blow would never have caused the person's death, as the dislocation might have remained unreduced without any danger to life. A coroner's inquest is held in every instance of death during the administration of chloroform, and, so long as the administrator can show, not only that he had no reason to think the patient could not take an anæsthetic safely, but that the administration was conducted with due care and skill, the result of the inquiry will be to exonerate him from all blame.

Self-inflicted wounds.

Imputed or self-inflicted Wounds.—Various causes may explain the occurrence of self-inflicted wounds; in women they are often to be ascribed to hysteria; in men they are usually the result of an intention to bring a false charge against someone, and so perhaps extort money; or they may, as in the following case, be given with the view of concealing a crime, and giving rise to the false belief of an attempted murder. A man was found lying in a room, which had been set on fire, by someone else according to his statement, and near him was the body of the deceased, who had evidently been killed by violence, the skull being fractured. When first found this man (the prisoner) was either insensible, or pretended to be. He stated that he had been suddenly attacked by a man, and knocked down by a blow on the right temple; when he got up he was again knocked down, and he then felt a knife at his throat, but it did not occur to him to put his hands up to protect his throat; his hands were not injured. He remembered receiving some blows on his body, and he then became insensible. There was a wound in his throat about an inch and a half long on the left side below the jaw. It only went through the skin, and a very little blood had escaped from this on to his cravat. There were many cuts on his coat at the back and sides, through his waistcoat, shirt, and flannel shirt, but there were no corresponding stabs or cuts, or indeed any marks upon his skin. The question then was whether these wounds were inflicted by the imaginary assailant, who had murdered the deceased and set fire to the premises, or whether the prisoner had committed the murder, and inflicted the wounds upon himself, and then set fire to the premises to conceal his crime, and give an air of plausibility to this version of the affair. No motive could be discovered for the commission of this crime by the prisoner, and he had borne a good character previously, but the wounds on his own body were so clearly self-inflicted that he was convicted of manslaughter.

These self-inflicted wounds, not given with any intention of committing suicide, generally have this in common, that they are so managed as not to do much harm; wounds likely to be fatal are not inflicted. As a rule, cases of this description should present little difficulty; the wounds in the clothes and on the body do not by any means harmonise, and when this is the case suspicion of imposture should be at once aroused.

CHAPTER XII.

WOUNDS OF DIFFERENT REGIONS OF THE BODY.

It will be advisable to deal with wounds of the different parts of the body a little more in detail, bearing in mind that the object is to determine the relative danger of such wounds.

Wounds of the Head.—Scalp wounds, incised or punctured, are amongst the commonest forms of injury that surgeons are called upon to treat, and, in the vast majority of instances, they get well without any ill effects, but occasionally erysipelas sets in, or sometimes, at the end of perhaps a week, secondary hæmorrhage appears, and proves very troublesome : the onset of this may sometimes be suspected when a person, about the fourth or fifth day, complains of great pain in the wound, and there is nothing to be seen. But the danger in cases of injury to the head is of damage to the brain. In every case of scalp wound, no matter how slight it may appear, the surgeon should probe the wound carefully to feel if the bone is exposed anywhere, or if any portion is depressed; without taking this precaution it is impossible to estimate the amount of damage that has been done, and even though nothing wrong can be made out on such examination, there is always the possibility of fracture of the inner plate of the skull-cap.

Symptoms do not always come on immediately when there is a fracture of the skull. Thus there is a case on record where a man fell from a scaffolding and was taken to a hospital; on arriving there he was able to walk upstairs and talk about the accident, and yet the post-mortem examination, three days later, showed that his skull had been split in half from before backwards, the longitudinal sinus being laid open in its whole length. Concussion of the brain is the name given to the group of symptoms, which may succeed a fall or blow on the head, where no injury can be recognised with the naked eye. In its most severe form death may be almost instantaneous; from this there are all degrees of severity, down to mere faintness. Few, if any, problems, in the whole range of medicine, present greater difficulty than those connected with this subject. To determine whether a person is suffering from the effects of intoxication, or concussion, or of a mixture of these, and if so, in what proportion, is a matter of the utmost difficulty in many cases, and the frequent occurrence of a paragraph in the newspapers, headed 'Drunk or Dying?' testifies to the number of mistakes that are made in the endeavour to discriminate between them. In the absence of external signs of cranial injury there are no signs upon which reliance could be placed; inequality of the pupils should always be regarded as pointing to cerebral mischief, and in any case of doubt it would be proper to give the patient the benefit of the doubt. To treat a drunken man as if he were suffering from concussion would not do him any harm, whereas to treat a person suffering from concussion, in the heroic manner often necessary to rouse a person from a drunken stupor, could but tend to render his death, which before was only a possibility, a certainty. The most common accident, perhaps, where no external damage is done in fatal cases of head injury, is rupture of the middle meningeal artery (which grooves the internal plate of the temporal bone); the

Cerebral injuries.

blood in such a case is poured out with some rapidity at first, and causes a compression of the brain giving rise to anomalous cerebral symptoms, and, as a rule, causing the patient's death in the course of a day or so. Fracture of the base of the skull is one of the most serious accidents that can happen, but few cases of recovery having been recorded. When the fracture passes through the petrous portion of the temporal bone there will be a discharge of cerebro-spinal fluid from the ear, and when it passes through the ethmoid bone there will be a similar discharge through the nose. In the absence of both of these signs the condition may be surmised, but a positive diagnosis cannot be given. In old people, or those whose blood-vessels are diseased, as in the case of Bright's disease, mere excitement or passion may suffice to rupture a vessel, and this spontaneous rupture not infrequently takes place soon after a meal. Great confusion has sometimes arisen from the fact of an old man having a fit after some trifling blow or accident, and on post-mortem examination being found to have died from apoplexy. That a blow might cause the rupture of an atheromatous artery it would seem impossible to deny, and yet it would be at least as difficult to affirm that it actually did so in any particular instance.

A medical man must, however, be very careful not to be too dogmatic as to the impossibility of recovery after cranial injury. The celebrated American crowbar case, as it has been called, illustrates this: An iron bar, weighing $13\frac{1}{4}$ lbs. and measuring three feet seven inches in length, and an inch and a quarter in thickness, was driven by a gunpowder explosion through the skull of a man, who ultimately recovered, with the exception of the loss of one eye. The bar entered under the zygomatic arch and carried away nearly the whole of the frontal bone; a very considerable quantity of brain substance, from the anterior and temporo-sphenoidal lobe on the left side, must have been destroyed. The man is said to have been slightly convulsed at the time of the accident, but spoke in a few minutes; he rode in a cart away from the scene of the accident, and was able to sit up during the journey, a distance of three-quarters of a mile; at the end of it he was able to walk up a flight of stairs with assistance, and retained his senses and memory so as to be able to give a connected account of the occurrence. As already stated, he made a good recovery, the only change, besides the loss of his eye, being a marked alteration in his temper, a point of much interest in relation to the question of cerebral localisation. Had the injury extended farther back, and damaged the brain in the neighbourhood of the fissure of Rolando, there can be no doubt, in the light of recent advances in this subject, that permanent hemiplegia, of the opposite side of the body, would have resulted. The case is very important, as showing that shock is not a necessary accompaniment of an accident, of the most serious nature, to the brain.

Wounds of the Face.—Beyond the disfigurement they lead to and the risk of erysipelas, these are not of much moment, when the lower part of the face is the part wounded, but the frontal bone is very easily fractured, and the brain damaged in that way: cases of epilepsy have been known to follow a comparatively slight blow on the frontal bone. The orbit, however, is the part of the face most obnoxious to wounds; penetrating wounds are easily given in this region, and the consequences are very serious. In one case a knitting-needle pushed into the orbit, in a lover's somewhat rough play, passed by the eyeball, without damaging the eye, deeply into the brain, and was followed by hemiplegia, of the opposite side of the body, with rigidity of the paralysed limbs.

<small>Injuries to the face.</small>

Wounds of the Neck.—Fracture of the larynx, the result of great manual violence, is always homicidal, and is a very dangerous accident, setting up an obstruction to respiration, and, even when tracheotomy has been performed, has often proved fatal. Incised wounds of the trachea, when the other structures in the neck have not been wounded, do not, as a rule, prove fatal, but wounds of the œsophagus are much more dangerous to life. The chief danger of wounds in the neck is of course that of hæmorrhage, from one or more of the numerous vessels in this situation. Death does not, however, always result immediately, even when all the vessels have been divided: thus in a case of suicide, a man lived for half an hour after all the structures, on one side of his neck, had been divided down to the bone; and in another case, one of murder, the deceased was able to run 23 yards, and had climbed over a stile after receiving a wound in the neck, which divided the common carotid artery, several branches of the external carotid, and the jugular-veins.

Wounds of the Spine.—The danger of these lies in the damage done to the spinal cord. The most common forms of injury are fracture and dislocation, both nearly always the result of accident. Homicidal wounds in the form of stabs are occasionally met with, suicidal wounds in this region are almost unknown. Fractures or dislocations affecting the atlas or axis are, almost invariably, followed by speedy death, though a very few persons have survived the spontaneous fracture of this bone from disease. Any damage to the spinal cord, above the fourth cervical vertebra, will probably prove immediately fatal, owing to the effects either on the medulla, or, if below this, on the cord above the origin of the phrenic nerves, interference with whose function must be followed by more or less speedy death.

Injuries to the spinal cord below this point produce complete paralysis of motion and sensation of all the parts below, and when the injury is seated high up in the cord death almost always occurs soon, from the respiratory troubles induced by the paralysis of the chest-walls. If lower down, the patient may recover, but is, as a rule, crippled for life; inflammatory changes having been set up in the cord leaving the limbs paralysed, and the muscles in a state of contraction.

Since the introduction of railway travelling and railway accidents, a new form of injury has sprung up, namely, spinal concussion, often called the 'railway spine.' The exact cause of the symptoms is obscure, but it would seem that it is not necessary that the patient should have been thrown violently against the back of the carriage. The symptoms do not commence at once; the sufferer has often been able to walk away from the scene of the accident, and has been congratulating himself upon his lucky escape, when in the course of a few days he may begin to suffer from pain in the back, difficulty in walking, and sometimes difficulty as to his bladder and rectum. Death may happen in the course of a few days, with the symptoms usually associated with acute myelitis, or the patient may linger on for months developing all the symptoms of chronic myelitis or pachymeningitis, including atrophy of the optic discs, a valuable sign from a diagnostic point of view. Unfortunately there can be no doubt that a great many of the cases of so-called spinal concussion are pure impostures, for the sake of extorting money out of the railway companies, and when a verdict has been obtained in his favour, only too often the would-be cripple is speedily restored to health. The symptoms are so easily feigned, and the diagnosis rests so much on the patient's own statements, that deception is

Railway injuries.

rendered easy, and its exposure a matter of considerable difficulty. It would appear that in the majority of cases, in which claims for personal damage have been made, the injuries were not real; this will tell very hardly upon those whose injuries are *bonâ fide*, for juries will be less liable to award substantial damages, when they learn that in the majority of cases the symptoms are feigned.

CHAPTER XIII.

WOUNDS OF THE DIFFERENT REGIONS OF THE BODY (*continued*).

Wounds of the Chest.—When these affect the chest-wall they are not, as a rule, dangerous; the intercostal arteries are well protected by the ribs, and are only likely to be injured by a thrust from below upwards. Hæmorrhage from them, or from the internal mammary arteries, may be difficult to arrest. Fracture of one or more ribs is, as a rule, not a very serious accident, but if the lung be wounded there may be serious hæmorrhage from it, or there may be extravasation of air into the pleura, and into the subcutaneous tissues of the whole body. Pneumothorax thus caused may sometimes prove fatal. Wounds which penetrate the chest are always of a dangerous nature, as, in addition to the lungs or pleuræ, the heart and pericardium and the great vessels may be injured, or if the wound reaches very deeply, the œsophagus, thoracic duct, or sympathetic nervous system. A wound of the lung may be recognised by the person spitting up blood; if the hæmorrhage take place into the pleura, hæmothorax is formed, and the lung on that side may rapidly and completely be compressed. *Penetrating wounds of the chest.*

Wounds of the heart, as a rule, produce very speedy death, as the blood is poured out into the pericardium, and the heart's action thus mechani- *Wounds of the heart.*

cally impeded. There are, however, on record some cases showing that even the most serious injuries to the heart are not of necessity immediately fatal. The following case, from Taylor, illustrates this: An Italian, aged 38, discharged a brace of pistols into his chest on the left side. He was brought to the hospital, was able to converse on his condition, and lived one hour and fifteen minutes. After death it was found that one bullet had entered the right ventricle, and, after perforating the septum ventriculorum, had made its exit from the heart at the junction of the left auricle with the ventricle. It traversed the upper lobe of the left lung, and was found fixed in one of the vertebræ. The second bullet perforated the left ventricle, and then traversed the left lung; the wound in the ventricle was of such a nature that at every contraction the opening must have been closed, and so the flow of blood was prevented. This man, owing to severe suffering, rolled about on the floor, and was with difficulty kept quiet. It will be seen that there were in this case bullet wounds traversing completely the cavities of the heart, yet the man could talk and exert himself, and survived their infliction for an hour and a quarter. In another instance a man survived a laceration of the left auricle for 11 hours. Rupture of the heart has sometimes been found in children, who have been run over in the street and killed, without any external sign of injury. As regards wounds of the great vessels, the following case is worth bearing in mind: A young woman during a scuffle received a wound in the chest from a shawl-pin. When taken to a hospital immediately afterwards, the pin was found to have pierced the chest close to the sternum, only the head being visible; her breathing was much distressed. After some deliberation the house surgeon decided to withdraw the pin, and accordingly did so, and the woman died almost immediately. At the post-mortem examination, it was found that the aorta had been transfixed just at its commencement, inside the pericardium; when the pin was withdrawn the pericardium was at once filled with blood, whence the sudden death of the patient.

Wounds of the Abdomen.—As has already been mentioned, a blow on the abdomen in the epigastrium may be followed by sudden death, and as yet no proof has been obtained of the exact cause of death in these cases. The epigastric artery may be injured by a wound on the lower part of the abdomen, which might give rise to a serious or even fatal hæmorrhage. The chief danger, from punctured wounds of the abdomen, is the injury to the peritonæum, and subsequent peritonitis, though with modern methods of dressing such wounds it is probable that, if seen sufficiently early, there would be a fair prospect of recovery. The following case shows that all penetrating wounds of the abdomen are not fatal: A soldier, running away after trying to commit a crime, was received by the sentry in the position of 'charge bayonets,' and being unable to stop himself, he ran on to the point of the bayonet, which entered his body an inch to the left of the ensiform cartilage, and passing through the abdomen emerged, near the vertebral column, some inches lower down. Two minutes later he was seen by the surgeon of the regiment—the bayonet having meantime been withdrawn—seated at the table quite unconcernedly, and a quarter of an hour later he marched three-quarters of a mile to the hospital whence a fortnight later he was discharged. The day after his admission into the hospital there was a little blood in the urine, and subsequently there was a general anæsthesia of the walls of the thorax and abdomen, which lasted for a little while, but otherwise there were no symptoms, and his recovery

Abdominal injuries.

was complete. Other cases almost as remarkable might be quoted. Any of the abdominal viscera may be ruptured by external violence without a wound; in railway accidents, or when people are run over, rupture of the liver, spleen, or kidneys is not uncommon. In the case of the two former the accident is probably always fatal, whilst there is reason to believe that some cases of recovery after rupture of the kidney have occurred; but, as in many other things, it is not easy to diagnose rupture of any of these viscera, the shock and pallor, pointing to internal hæmorrhage, being almost the only guides. The presence of blood in the urine does not necessarily imply rupture of the kidney. Rupture of the intestines, which is excessively rare, must always be fatal in the absence of surgical interference. Rupture of the bladder may result from a kick on the lower part of the abdomen, but it can only happen when the bladder is full. It is, as a rule, followed by immediate symptoms; but a case has been put on record, in which a man walked two miles after sustaining a rupture of his bladder. The immediate result of this accident is, of course, extravasation of urine into the pelvis, and perhaps into the peritonæum, depending on whether or not the rupture is wholly extra-peritonæal. Wounds of the diaphragm, though very serious, are not always fatal. It is said that they never heal, so that the contents of one cavity are liable to protrude into the other, and there is a danger of the stomach or intestines becoming strangulated at the opening.

Rupture of the bladder.

Wounds of the Genital Organs.—In the female such wounds are almost always the result of violence by others, and, as the parts are very vascular, there is no inconsiderable risk of death from hæmorrhage. Wounds of the male genital organs, on the other hand, afford a strong presumption of being self-inflicted, and, it may be said, of

Injuries of the genital organs.

insanity also. Within the last few years, in this country, a case occurred, where a man accused two others of inflicting wounds of this nature upon him, and two years later he was again found with similar wounds; ultimately, on his death-bed, confessed that on each occasion they had been self-inflicted, simply to trump up a charge against two men towards whom he bore a grudge.

Fractures and Dislocations.—In the case of fracture of the skull, it may be asked, whether the fracture was likely to have been produced by a given weapon. This is a question that will not be easy to answer. In a few cases there may be something peculiar about the shape of the wound, which would suggest the probability of its having been caused by a blow, rather than a fall, but very often this question could not be answered. A fracture must not always be taken as an evidence of great violence; sometimes bones are abnormally thin or unusually brittle. This is especially the case in certain diseases of the nervous system—*e.g.* locomotor ataxy, where spontaneous fracture of the femur, or some other long bone, is quite common. It is probably owing to abnormal fragility of the bone, that the ribs are so often fractured in lunatic asylums, as there can be little doubt, that the ribs of the insane have been found fractured under circumstances which would not have caused them to break in healthy individuals. A dislocation produces at the time of its occurrence a complete disablement of the limb, a fracture does not always do so. A joint, that has once been dislocated, will be liable to be dislocated on a repetition of the cause, or even by a slighter cause.

[marginal note: Spontaneous fractures.]

CHAPTER XIV.

GUNSHOT WOUNDS, CICATRICES, BURNS AND SCALDS.

Characters of the wound. THE effects of gunshot wounds differ according to the weapon used, and the distance from which the shot was fired. A rifle bullet coming from a little distance and with considerable velocity will make a small clean-cut wound, not unlike a punctured wound, and apparently a good deal smaller than the bullet which passed through, whereas the same bullet, if fired from quite close, would make a contused and lacerated wound. When a shot gun is fired from a short distance the shot all penetrate together making only one wound, which would have all the characters of a lacerated wound; if fired from a little distance the shot would have had time to spread, and there would then be no mistake as to what kind of weapon had been used. When the bullet passes through the clothes some portion of these will usually be found imbedded in the wound, and if the shot was fired from near at hand portions of the wadding might be found in the edges of the wound, and grains of powder would be buried in the clothes, or surrounding skin, as the case might be. In the absence of these two latter signs it would be probable that the shot must have been fired at such a distance as to exclude the idea of suicide, unless the pistol, or other weapon, were found firmly grasped in the hand of the deceased; this would outweigh almost all other evidence against suicide.

If the bullet have passed through the limb or other part of the body, there will be a wound of exit as well as one of entrance; the former is always more or less ragged, and its edges are everted. According to Casper, the wound of entrance is always larger than that of exit, whilst other authorities state that the reverse is the case. It is probable that their relative size depends entirely upon the distance from which the bullet was fired; if from very near, there can be little doubt that the wound of entrance might be larger than that of exit; but if the ball came from a distance with a considerable velocity, it would appear to be certain that the entrance wound would be relatively small. A bullet may split on a bone and so make two wounds of exit, or it may enter the body a second time, and so there might be two wounds of entrance, though only one shot had been fired. Taylor refers to a case where one bullet caused two wounds of entrance, and three of exit, by passing through both legs in the way described. A bullet may sometimes be turned from its course by a very slight cause, and cases have been known where a bullet has almost encircled a limb instead of passing straight through. One of the most remarkable of such cases, perhaps, was that where a bullet entering the neck, and striking the larynx obliquely, passed completely round the neck under the skin, and lodged close to the surface on the other side of the larynx, whence it was easily removed.

Wounds of entrance and exit.

It must not be forgotten that it may not be possible to determine, from the character of the wound of entrance, what kind of missile was fired. Wadding, or paper, or gunpowder, if fired from quite close, will produce a wound which might, at

146 MEDICAL JURISPRUDENCE.

any rate on a casual examination, be supposed to have been produced by a bullet.

Examination of the weapon. The question may be asked whether a given weapon has been fired recently. The following are the steps recommended to be taken in order to answer this question: Wash the barrel thoroughly, after noticing the naked-eye appearances of the contents, with distilled water, and filter the washings. Test the filtrate (1) for sulphuric acid by adding a salt of barium; (2) for alkaline sulphides by adding a lead solution; and (3) for iron salts by ferro- or ferri-cyanide of potassium. If the bore of the barrel has a bluish-black colour, and contains no green crystals of ferrous sulphate and no rust, but the solution has a pale yellowish colour, smells of sulphuretted hydrogen, and gives a black precipitate with acetate of lead, this indicates that not more than two hours have elapsed since the weapon was discharged. If the barrel is less dark and no rust or crystals are found, but there are faint traces of sulphuric acid, then more than two but less than twenty-four hours have elapsed. If there are numerous spots of rust in the barrel, and if there are very clear reactions of iron in the rinsings, then at least twenty-four hours have elapsed, and perhaps five or six days. The longer the time that has elapsed the more crystals of ferrous sulphate would be obtained. If there is a good deal of rust, but there are no iron reactions in the rinsings, it has been discharged at least ten days but not more than fifty. It is not possible to go farther than this with any reasonable degree of probability.

Cicatrices. In all cases of disputed identity, the existence of scars will occupy an important place. It must be borne in mind, however, that two persons may have a scar in exactly the same situation, and this may be so even where there are several scars, so that their absence is of more value in evidence than their presence. The identification of the body of Harriet Lane, who was murdered by Wainwright, rested almost solely on the presence of a scar on one leg, which had been produced many years before death, and which had so far resisted decomposition as to be readily recognised, when the body was examined a year after the murder. This incidentally shows the great power scars have in resisting decomposition. Every wound, which passes through the true skin, will be followed by a scar, the extent of which will depend upon the nature of the original wound, and the length of time occupied in the healing process. When the wound first closes the resulting scar is reddish, and retains this colour for a week or so, it then becomes more brown, and in the course of a few months it assumes the dead-white shiny appearance, which it retains ever after. Cicatricial tissue is more dense than the ordinary skin, and contains less blood; this can be readily demonstrated by compressing the skin round a scar and the scar itself; on removing the pressure the surrounding skin will be immediately reddened by the flow of blood to it, and the contrast with the white scar will be more marked than ever. Cicatricial tissue is contractile, so that a scar is smaller than the wound which preceded it; it would not be wise to attempt to determine the size of the weapon with which a wound had been inflicted merely from an examination of the scar, at some period long subsequent to the original injury. The generally accepted opinion in this country respecting scars is that they are indelible, and that no lapse of time will suffice to remove all traces of a scar once formed. This belief formed part of the most conclusive evidence against the claimant to the Tichborne Estates. The real Roger Tichborne, supposed to have been drowned twenty years before the last

The Tichborne case.

trial, had a scar in one eyelid from a fish bone, that got caught in it, and had to be drawn through; also a scar on his left shoulder, where an issue had been kept open two years, and he must also have had scars of repeated bleedings on the arms, and of single lancet cuts on the temple (he was bled from the temporal artery once, as well as repeatedly from the arms). The Claimant had none of these scars, or rather had no scars in any of these situations, thus affording in his own person the strongest presumption, if not absolute proof, on medical grounds, that he was not the person he represented himself to be.

Scars from disease.

Scars may result from disease as well as from wounds, and all ulcers must leave scars behind them. Strumous ulcers leave scars irregular in shape with thickened margins, whereas syphilitic scars are notable by their great loss of substance. The scars left from burns and scalds are generally large and superficial, and are apt to lead to very unsightly deformities by the contraction to which they give rise. It would, however, be the situation, as much as the characters of the scar, that would be relied upon in pronouncing it to be strumous or syphilitic. Another form of scar requires to be mentioned, that due to overstretching of the skin; on the abdomen of a woman, who has borne a child or children, there will be found radiating white silvery lines, known as the lineæ albicantes, which afford very good, though not conclusive, evidence of pregnancy at some anterior period. A scar can never be entirely obliterated, though by excising it a scar of a different shape might be substituted; still it would occupy the same region as the original one, though perhaps it would not be of the same extent. The marks left by vaccination may totally disappear, but those from smallpox are, as a rule, indelible.

It will probably be more generally admitted that tattoo-marks may disappear than that scars can. A great deal will depend upon the kind of ink or colouring-matter used, and the person who performs the tattooing. If done with common ink, blue ink, or cinnabar, and by someone who had not any experience in the matter, it is quite possible that, in the course of time, such tattooing might become completely effaced; but if Indian ink, gunpowder, coal-dust, soot, or washing-blue had been used, and the tattooing had been efficiently done by an expert, it is very doubtful if all traces would ever be removed. This question also was raised at the Tichborne trial, the missing man having been extensively tattooed on his arm with a heart, cross, anchor, and the initials R. C. T. The Claimant had none of these tattoo-marks upon him, nor had he any scars in any of the situations where these should have been; but he had—which was very suspicious—a scar on his wrist in the exact situation where Arthur Orton had been tattooed with the letters A. O. So that the evidence on this ground was as completely against the Claimant as that in the matter of scars. Of course tattoo-marks may be got rid of by means of caustics, or by excision, but then a scar would be left, which would have to be accounted for. Some of the colouring matter will usually be found in the lymphatic glands nearest the tattoo-mark. *Tattooing.*

As *Burns and Scalds* act much in the same way, they do not call for separate consideration. Burns are usually divided into six classes, according to the severity of the burn, varying from the mere scorching of the skin to complete destruction of all the tissues. For the particular characters of each of these classes the student is referred to the text-books on surgery. *Burns and scalds.*

Burns are almost always accidental, hardly ever

the result of suicide; occasionally a murderer may have recourse to fire to conceal the traces of his crime, and this fact necessitates a part of the present inquiry, viz. whether the burns were inflicted during life or not. In the case of a fire in a house many will die of suffocation before they are burnt, and others may die of shock or of fright. The danger of a burn, or scald, bears an exact ratio to the extent of surface of skin involved, and it is therefore this point that should be looked to in forming a prognosis, and not the depth of the burn, which is of secondary importance.

In endeavouring to determine whether a burn was inflicted during life there are two signs upon which a great deal of stress is always laid. The first of these is the presence of a zone of redness separating the burnt skin from the healthy; and the second is the presence of blisters on the burnt part. It is obvious that when the body is completely charred neither of these tests will be applicable, and no opinion could then be formed upon the point. It is not quite certain how long a time is required for the production of the redness, so that it is possible, in a feeble person dying very soon after the burn, that there might not have been time for its production, though, if the burn were the cause of death, this would seem to be unlikely. On the other hand, accidental experiments have shown that it is possible to obtain the formation of the redness round the burnt part, even when the burn takes place after death, provided it occurs soon enough, that is to say, before all vital action has ceased. Tidy sums up the evidence respecting the redness as follows: 'Although its absence may not be an absolute proof that a burn was not inflicted during life, nor its presence that it was, nevertheless it constitutes the strongest possible presumptive evidence as to the period when the

A zone of redness.

burn occurred, its presence undoubtedly indicating, even supposing the burn was not inflicted during bodily life, that it was at most inflicted within a few minutes of that bodily life, and at any rate during the molecular life of the part or tissue concerned.'

As regards the second point, the formation of blisters, it cannot be assumed, when there are none, that the burn was not inflicted during life. The blister is due to the collection of blood serum beneath the cuticle. Numerous experiments have been performed on the dead body with a view to settle the question as to whether blisters could be so produced. Dr. Tidy and the late Dr. Woodman conducted a series of investigations on this point, and arrived at the following conclusions : ' 1. That in some cases blisters may be produced after death by the application either of a flame (this being the easiest method of all), or of boiling oil, or of molten lead to the skin, but that in no case were we able to produce a post-mortem blister by the application of boiling-water. 2. That in no case were we successful in producing a blister on an amputated limb at a later period than thirty minutes after its removal, probably owing to the rapidity with which cooling takes place. In the case of dead bodies, however, excluding those that died of dropsy, vesications were formed in two cases as late as eighteen hours, and in several cases as late as twelve hours after death. 3. That the majority of these post-mortem blisters were found to contain no fluid whatever, the raised cuticle rapidly collapsing after the source of heat was removed. In some few exceptional cases there was observed under the cuticle a small quantity of thin watery fluid, containing a mere trace of albumen only. 4. That in the bodies of persons who had died with well-marked dropsy, serous blisters could always be produced, and at almost any period after death;

Formation of blisters.

but the serum found under such circumstances was invariably thin and watery, rarely tinged with blood, and contained but a trace of albumen. 5. That in no circumstances were we able in a dropsical subject to produce a blister by the simple application of boiling-water, unless a stream of boiling-water was poured for some minutes upon one spot. Even under these circumstances we were not always successful. 6. That in no case of a post-mortem blister, whether on dropsical bodies or on the bodies of persons who had died by accident, instantaneously or suddenly, have we ever noticed the slightest indication of a line of redness, nor of a reddened cutis vera after the removal of the cuticle. 7. Our experiments entirely confirm the observations of Chambert that the serum exuded in a blister formed during life is thick and rich in albumen. We ought to note that in badly-nourished people blisters may be found on the trunk and extremities, also that if a burn be inflicted on a person in a state of insensibility where the vital powers are much depressed, no line of redness or vesication may result, but should the person recover, blisters and a red line would become apparent; and lastly, that bullæ frequently form on a dead body as the result of putrefaction.'

Tidy sums up the evidence in the two following propositions: '1. Given a burn on a dead body where there are serous blisters, the serum being thick and rich in albumen, and the blisters surrounded by a deeply-injected red line, the true skin after removal of the cuticle also presenting a reddened appearance, the evidence is strong that the burn was produced during the life of the person, whilst it is conclusive that it was caused during the life of the part. 2. Given a blister containing air, the true skin after the removal of the cuticle appearing dry and unglazed, of a dull white colour, or dotted over with grey specks at the opening of the sudoriferous ducts; or given a blister containing a little thin non-albuminous serum, there being in neither case any red line surrounding the blister, nor any injected condition of the cutis vera or subcutaneous tissues, the evidence is strong that the burn was inflicted after death.'

Spontaneous Combustion. — Cases of alleged spontaneous combustion have been few and far between in this country, and have not been received with much credence when they have been reported, but in France they have been accepted by even leading medical authorities. Most of the instances, in which death has been attributed to spontaneous combustion, have occurred in women, and usually in corpulent people addicted to habits of intemperance. It seems to be believed that, under such circumstances a much slighter cause than would ordinarily suffice, such as a spark from a pipe, has set the body on fire. In these cases, too, the surrounding objects have not suffered from the flames, in anything like the proportion that would have been expected under ordinary circumstances.

CHAPTER XV.

LIFE INSURANCE.

Definition of terms. An insurance of a life is a contract between two parties, whereby one party undertakes, in consideration of a certain sum of money paid annually, and called the 'premium,' to pay to the representatives of the other, on his death, a certain sum of money. Or the contract may only apply to the death of the contracting party within a certain number of years. The deed, by which such contract is made, is called a 'policy.' In the case of an ordinary policy, payable at death, the money cannot be recovered until distinct proof has been given of the death of the person insured. The amount of premium payable will be proportionate to the amount insured for, the expectation of life of the particular person being taken into consideration. The method best known, for determining the expectation of life, is that called the method of Willick, and it consists in subtracting the age of the person from 80; two-thirds of the number so obtained will give the mean expectation of life of that individual in *Sound lives.* years. Thus, supposing a man to be of the age of 29 when he applies, subtracting this from 80 gives 51, two-thirds of which amount to 34; therefore this person has, from the insurance point of view, 34 years to live—in fact till he is 63; and if he should insure his life and die before that period the office would lose; whereas, if he lived beyond the age of 63, the office would gain.

These figures presuppose, that the person can show a clean bill of health, has a good family history so far as inherited tendency to disease is concerned, and is not engaged in any occupation likely to be prejudicial to his health. Some insurance offices only accept what are called first-class lives such as these, others take unsound lives, but at an increased premium.

In such a case the insurance offices, instead of issuing the policy at the ordinary rate, add on a number of years to the man's age, and charge him accordingly. For instance, supposing that the family history pointed to some grave disease, as likely to make its appearance, the office, instead of accepting the man at his real age of 29, as above, might offer to issue a policy to him on the hypothesis that he was 49; so that, instead of an expectation of life of 34 years, there would only be one of 21, and the annual premium, that he would be called upon to pay, would be half as much again as in the other instance.

Amongst the many causes of unsound lives may be enumerated the following: Those who show an inherited tendency to disease, whether tuberculosis, cancer, gout, dropsy, heart disease, or rheumatism. Next, those whose habits have been likely to permanently injure their constitutions, such as drunkards, or those who take opium. Lastly, those whose occupation may be considered to be, either unhealthy, or attended with increased risk to life; amongst these are included all engaged in sedentary occupations, especially if they work in badly-ventilated rooms, stone-masons, millers, and others whose occupation exposes them to the inhalation of finely-divided particles; also those engaged in blasting operations, and there-

Unsound lives.

fore liable to accidents from gunpowder; policemen, firemen, sailors, and soldiers. As regards soldiers, an additional premium is required, beyond and above the special rate, if they go to foreign climates. In women child-bearing greatly increases the risk that is incurred in insuring the life, although in itself child-bearing is but rarely fatal.

When a person wishes to insure his life he has to fill up a most exhaustive form as regards his own past illnesses, and the health of his brothers and sisters as well as of his parents, and the cause of death of any of these, supposing them to have died. He is also asked as to his own habits of life, and as to whether he has had any of certain definite symptoms, such as blood-spitting and others; and he is also obliged to answer a question as to whether he is, at the time of proposing to insure, and has been always, of temperate habits. He also has to give a reference to two persons who have known him, to whom the office apply for a confidential report. He also has to state whether he is insured in any other office, and whether he has ever been refused by any office, and if so, for what reason. He will also have to send in a report from his usual medical adviser, and, in addition to that, he will be examined by a medical man on behalf of the office.

Material concealment.

If there is any material concealment of facts, such as the insurers have a right to know, which would alter the risk that they are incurring in accepting the life of any person, the policy will be void. The medical man, who is called upon to sign one of these certificates, should remember that, if he undertakes 'it, he must answer every question to the best of his knowledge; it is no uncommon thing for men to leave unanswered the question about the habits of the proposed subject for insurance, simply because they know nothing about them. Such a course is mistaken and may do harm to the insured, as there is always a certain amount of suspicion attached to an unanswered question. If nothing is known as to the habits of a person, it is clearly the duty of the medical man to say so.

One not uncommon trick amongst the dishonest, who want to get insured, and who know that they have not got sound lives, from the point of view of the insurance office, is to call in a medical man who is more or less of a stranger to them, perhaps for some trifling ailment, and then afterwards refer to him as their regular medical attendant, suppressing the fact of serious illnesses, in which they have been attended by some other practitioner. For instance, a colonel in the army insured his life by two policies in May and June, and died of a remittent fever in the same year. The company refused to pay on the ground of misrepresentation and concealment. The gentleman who gave the certificate had not attended him for three years, and the applicant stated that he had never had any other medical attendant, and no serious illness. It appeared, however, that, in February and April of that same year, he had been attended by two other practitioners for fever with cerebral symptoms. Although his death had not in any way been dependent upon that illness, yet the policy was considered void, because there had been a material concealment of serious illness.

One of the subjects upon which, more often perhaps than upon any other, a payment of a policy is refused, is on the ground of concealed habits of intemperance. On this subject there is a very real difficulty, inasmuch as there is no standard, by which to determine what are habits of intemperance, one witness regarding the slightest staggering in the gait as conclusive of intoxication, another not considering a man to be drunk until he

Intemperance.

is unable to stand. When, however, an insurance company refuses to pay the amount of the policy, on the ground that the deceased had concealed from them his habits of intemperance, the *onus probandi* is thrown upon them. In a trial which took place in December 1884, it was laid down very clearly by the judge, and admitted by the counsel for the company, that it was incumbent on the company to prove (if they were successfully to resist the claim brought against them) that the deceased was habitually intemperate. In the particular case in question a great deal of contradictory evidence was given, and after an eight days' trial, the jury, being unable to agree, were discharged without a verdict.

As regards delirium tremens, the following case of concealment is worth quoting, as illustrating the difficulties, a medical man has to contend with, in finding out the past existence of such a condition. Payment was refused by the company on the ground that the written answers, made by the deceased to questions proposed by the company, were false. One question was whether he was subject to delirium tremens, or any disease tending to shorten life, which he answered in the negative; a second was whether he was of temperate and sober habits, which he answered in the affirmative; and a third was as to the name and residence of his ordinary medical attendant to be referred to regarding present and general state of health, to which he replied giving the name of a medical man. He died about two years after effecting the insurance. Payment of the policy being disputed by the office, the case came into court, when it appeared in evidence that the deceased had had an attack of delirium tremens within a month of effecting his insurance, and that the doctor, whose name he had given, had not attended him for three years, he having since then been attended by another medical man, to whom he had given no reference. An interesting point in the case was that this gentleman, his own real medical man, appeared as a witness for the office, and deposed to having attended the deceased during two attacks of delirium tremens, which were only a fortnight apart, whilst the regular medical adviser of the office, who had examined him on behalf of the office in the interval between these attacks, described him as a first-class life; he said he observed no indications of delirium tremens about him or of drunken habits, and stated that he presented the picture of health. The concealment of habits of intemperance was held to be clearly proved, and a verdict was returned for the insurance office.

The insurers must take the risk of the insured developing habits of intemperance after the policy is made out, their only ground of refusing to pay, in such cases, being that there was wilful misrepresentation at the time of effecting the insurance. It need not necessarily follow that the insured should have died as a result of the habits, which have been concealed, the mere concealment is sufficient to render the policy void. Thus, in one case, the habit of opium eating, on the part of the insured, was not mentioned, and when, many years afterwards, the insured died, this fact, of the concealment of the opium eating, was held sufficient ground for declaring the policy void, though there was no contention, on the part of the insurance company, that the death had in any way been caused, or accelerated, by the habit. Amongst other habits, which might exercise some influence on the duration of life, may be mentioned tobacco smoking, which, however, does not appear to be specifically inquired into by insurance offices, and yet there can be but little doubt that, when carried to an excess, as it often is, it must exercise some deleterious effect on the system. When, however, no question is asked, Lord Mansfield's remark may be borne in mind and acted upon: 'The insured need not mention what the insurer ought to know, what he takes upon himself the knowledge of, or what he waives being informed of; the insurer need not be told general topics of conversation.'

Any misstatement as to age, place of residence, or occupation, would certainly vitiate the policy. It must be noted, too, that the insured may have answered the questions to the best of his knowledge, but may have made misstatements nevertheless; if, however, there was no intention to deceive, the probability is that the office would have to pay. Men often do not know what illnesses they have had, and they may have had a serious malady without being aware of its grave nature. This remark especially applies to syphilis, which many persons contract without being aware of the fact at all, the immediate effects having passed away almost unnoticed. It could not be said that there had been any concealment of a disease, the existence of which was not known to the insured. The offices lay an amount of stress upon blood spitting, that is quite unnecessary as a symptom of pulmonary consumption, for the term is too vague to be regarded as of any value in a medical sense.

As to *suicide*, the majority of offices have a clause in the policy to the effect, that it will be void if the insured commit suicide, such suicide, of course, being not unintentional. If the death was the result of unintentional suicide, those who claimed the payment of the policy would have to prove that this was so. As a rule, all cases of suicide are investigated by a coroner's jury, but the insurance offices are not bound to accept the finding of that jury; for instance, if the latter find, as they mostly do, that the deceased committed suicide whilst labouring under temporary insanity, the offices are not bound to pay the policies on the strength of that decision, and may hold an inquiry into the death on their own account, but the onus is then thrown upon them of proving that the deceased was not of unsound mind when he committed suicide. In regard to the question as to whether the deceased was of sound mind when he committed suicide, one of the judges, at the conclusion of a trial, made the following remarks: 'In my judgment, if death be the result of *disease*, whether affecting the *senses* or the *reason*, the insurance office is liable under this policy. The act which is not the act of a sane responsible creature, but is the result of any delusion or persuasion, whether physical, intellectual, or moral, is not the act of the man.' Tidy sums up the consideration of this important topic as follows: 'It appears to us that if a person is clearly proved to be of unsound mind, and in that condition commits suicide, it is fair and reasonable to regard the suicide as one of the results of a diseased and unsound mind, and not as an act which is the exercise of an intention, or in any respect whatsoever of felonious killing. It would, no doubt, be right to require that the onus of proof, that the man was insane at the time, should be thrown on those who would benefit by the death.'

A curious case in the United States occupied the attention of the medico-legal world for some time in 1883. Colonel Dwight had insured his life, for large sums, in most of the insurance offices throughout the country, having in fact been accepted or refused by all but one single office in California. After payment of one quarterly premium, and before the second became due, he died, being at the time of his death a bankrupt, and having

The Dwight case.

made no arrangements to pay the second instalment, which would have been payable in a few days. The symptoms of his illness were very obscure, and were supposed to be due to some form of malarial fever, for which he was treated by his medical men with unusually large doses of morphia. The post-mortem appearances revealed nothing definite; the heart was empty, the lungs showed a little superficial emphysema, and were otherwise congested, the liver, spleen, and kidneys were somewhat congested. On the neck there was a deep indentation, with a parchment-like base, broad enough to take a finger laid in it, not crossing the neck in front but extending all round the back of it. The exact mode of death was somewhat doubtful; it had not been expected by his physicians, who had left him an hour and a half before; and the account given of it by some of the witnesses was quite consistent with asphyxia. The undertaker and two witnesses swore that the mark was not on the neck when they saw the body after death, and they attributed it to the forcible flexing of the head forward on putting the body into an ice-box. The medical witnesses were almost unanimous, in attributing death to strangulation by means of a cord or rope, and in stating that no natural cause of death had been found. The verdict, however, was given against the insurance offices, that is to say, in favour of death having been due to natural causes. It may be added that the signs usually associated with an early stage of general paralysis of the insane were found in the brain, and that, in all probability, the monetary transactions of the deceased should be regarded as symptomatic of that disease.

It is a common thing, at the present day, for persons to insure against accident, and there are several offices which only deal with cases of accident. In such there is usually a scale of payments agreed upon, in case of temporary or permanent disablement, or a fixed sum in case of death. What may be held to constitute an accident is often the subject of dispute. Disease caused by the occupation, or the place of residence, or changes in the weather will not be accounted of accidental origin, and sunstroke has been held not to be included under the head of accidents. Whether death by lightning would be considered an accidental death is doubtful, though a person dying thus might, in ordinary language, certainly be described as having been accidentally killed. In the United States it was decided, in one case, that the insurance office was not bound to pay when death had been due to an accidental overdose of opium. A nice point has sometimes arisen, as to whether a person's death was to be attributed to accident or disease; it is quite conceivable that a person, labouring under some serious malady, might meet with an accident, which would have been of no great moment to a person in robust health, but from which he, in his weak state, is unable to recover, and the question may then arise whether the office could be compelled to pay the policy. The points, which would have to be taken into consideration, are thus concisely put by Tidy : 1. The state of the person's health before the accident. 2. The relationship in point of time between the accident and the setting in of the disease. 3. The connection between the special disease from which the person suffers and the accident stated to have occurred. *Death from accident.*

The law endeavours to prevent, as far as possible, one person having an interest in another's death by reason of an insurance, effected on his life, and any insurance effected on the life of another by anyone who cannot show that he has a direct legitimate interest in the life of such a person, is *ipso facto* void. The insurance money has been in

The evils of the insurance system. past times a fertile source of murder, the most notorious instances being the murders by Palmer in this country, and De la Pommerais in Paris. In order to render such cases impossible it would be necessary to prevent the sale, or assignment, of an interest in a life policy altogether, as the possessor of such a policy can only have an interest in the death of the insured at the earliest possible moment. On this subject it is necessary to refer to the objectionable system of infant burial clubs, which prevails so largely amongst the poor, and proves a most fruitful cause, it cannot be doubted, of infant mortality. The only way to check this evil would be to insist, that the burial clubs should pay the funeral expenses according to a fixed scale, and that no money should be directly received, by the parents, in respect of the death of a child.

Feigned death.

In concluding this subject, notice may be taken of the feigned deaths which have from time to time occurred. A man goes down to bathe and is never seen again, his clothes are found by the sea-shore, and the friends claim the insurance money. This has happened more than once in this country, and in one case, where the insurance office paid the claim, the man was seen in South America within two years of his disappearance. In America a more elaborate fraud was tried, and, for a time, successfully. The assured, and his associates, procured a dead body and placed it in an outhouse, to which they then set fire; the remains were then sworn to as those of the insured. The secret came out afterwards, owing to the insured not thinking he was being fairly treated by one of his associates. The latter murdered him, and being convicted of this crime, the previous fraud was discovered.

CHAPTER XVI.

VISION.

THE medico-legal aspects of vision refer to the limits of distinct vision in the normal eye, and the correct appreciation of colour. In almost every criminal trial the identification of the accused forms a most important part of the evidence. Those who are constantly accustomed to study faces attain a degree of power, in this respect, to which ordinary individuals can by no means approach; thus prison warders and detectives can in general recognise a man with whom they have once had to deal, no matter how many years have elapsed since they last saw him. In the case of burglary or murder, very often the light is insufficient, which adds to the difficulty of one inexperienced in observing faces with a view to recognition. A flash of lightning, or even the flash of a pistol, has sufficed to enable a person to see the features of a criminal so as to be able to swear to him afterwards, but such instances are rare.

In the army and navy (and in many occupations and professions) good vision is essential, and in both these services the acuteness of vision is tested, and, if not up to a certain standard, the candidate is rejected. But no tests are applied to ascertain whether the candidate possesses a normal colour vision, which, for the sailor at any rate, is quite as essential as a normal range of vision, and there can be no doubt that many of the, apparently inexplicable, collisions, which have happened at sea, have been owing to the fact that the officer on duty was unable to distinguish between the red and the green lights. In engine-drivers, too, it is easy to see how dangerous it would be for a man not to have a perfect colour vision, and yet no tests, or very incomplete ones, are adopted before men are employed in such a capacity.

CHAPTER XVII.

INSANITY.

No definition of insanity, which has yet been put forth, has met with universal acceptance. Insanity is disease of the brain affecting the mind; but this does not express the essential feature of insanity from a legal point of view, viz. that it is a disease rendering a man irresponsible. Dr. Bucknill has lately proposed a new definition, which fulfils this requirement, and, further, has the advantage of being concise. He says insanity 'is incapacitating weakness or derangement of mind caused by disease.' The question of capacity, or incapacity, is the practical point to be decided in every case of supposed insanity, and the medical jurist is only concerned with insanity in so far as this has to be determined. *Definition.*

There are many varieties of insanity as described by the College of Physicians, which has published the following classification:— *Classification.*

Mania.
Melancholia.
Dementia, including acquired.
Imbecility.
Idiocy, congenital imbecility.
General paralysis of the insane.
Puerperal insanity.
Epileptic insanity.
Insanity of puberty.
Climacteric insanity.

Senile insanity.
Toxic insanity, alcohol, gout, lead, &c.
Delirium tremens.
Traumatic insanity.
Insanity associated with obvious morbid change or changes in the brain.
Consecutive insanity, from fevers, visceral inflammations.

In other classifications the causes alone have been taken as a guide, and in others, again, the forms of the disease alone. The points, which have to be taken into consideration, are the relations of insanity to contracts and to wills, competency to give evidence, the imposition of restraint, and the relation of insanity to crime.

Contracts. The validity of a contract made by an idiot could not be upheld. In the case of an imbecile it would be a matter to be decided by evidence whether that particular imbecile was competent to make a contract. If a party entered into a contract with a lunatic, and if the lunacy of that person was of such a kind as to affect his capacity, then that contract would be voidable. The case would be otherwise, however, if the contracting party were ignorant of the incapacity of the lunatic. Thus, in a case of this kind the judge laid it down that 'where a person, apparently of sound mind, and not known to be otherwise, enters into a contract for the purchase of property, which is fair and *bonâ fide*, and which is executed and completed, and the property, the subject-matter of the contract, has been paid for and fully enjoyed, and cannot be destroyed, so as to put the parties *in statu quo*, such contract cannot afterwards be set aside, either by the alleged lunatic, or those who represent him.' Marriage contracts are amenable to exactly the same interpretation as ordinary contracts, and the insanity, which would void an ordinary contract, would also void a marriage. In such a case, however, if the insane party should subsequently be restored to reason, he or she, as the case may be, may ratify the marriage. If the person be only of weak mind, the marriage will not be voidable unless fraud, or undue influence, can be proved to have been used. This was seen in the case of the Earl of Portsmouth early in the present century. The earl

Avoidance through undue influence and fraud.

was known to be weak-minded, and, though of age and a widower, was in the hands of guardians. One of these, a solicitor, persuaded him to marry his daughter, without communicating with the relations or other guardians, and the marriage was afterwards declared void on account of the undue influence used.

There is, perhaps, no question in the whole range of medical jurisprudence which comes before the public more frequently than that of the competency of a person at the time of making a will. Almost every day the newspapers contain the record of a disputed will case. The capacity of the testator, as it is termed, has to be decided. Had he, at the time of making his will, a disposing mind? that is, one capable of independent comprehension. According to Taylor, a person is considered to be of disposing mind, who knows the nature of the act he is performing, and is fully aware of its consequences. An idiot has not a disposing mind, but if it could be shown that an idiot had conducted any business transaction with intelligence there would be *primâ facie* ground for believing that idiocy did not exist. The law, in this country, has never defined the exact degree of intelligence necessary to enable a person to execute a will. That is left to be decided by the evidence in each individual case. An eminent judge, speaking on this point a few years ago, said that if distinctions could be made between the different degrees of soundness of mind, the highest degree was required to make a will, because it involved the power on the part of the testator of reflecting on the relative claims of different people, and of determining in what proportions his property should be divided.

Mere weakness of mind would not in itself invalidate a will, though if undue influence or fraud were proved, the combination of the two would

marginalia: Wills. A disposing mind.

probably suffice to secure the cancelling of the will, as actually happened, in reference to a marriage contract, in the Portsmouth case. Sometimes a will is disputed on the ground that its contents afford evidence of insanity. But this line of argument seldom obtains much weight in the law courts. A testator must always be absolutely free to dispose of his property as he pleases, and mere caprice cannot be accounted insanity. The most difficult cases, of this class, are those in which a decision has to be made as to what is mere eccentricity, and what is something more than this, and, in fact, amounts to insanity.

Eccentricity not insanity.

There are other grounds on which the capacity of the testator might be called in question, viz. cases in which he had reached old age, or was dangerously ill at the time of making the will. Mere old age in itself does not afford any presumption of incapacity on the part of the testator. 'It is admitted,' said Lord Chief Justice Cockburn, ' on all hands that though mental power may be reduced below the ordinary standard, yet if there be sufficient intelligence to understand and appreciate the testamentary act in its different bearings, the power to make a will remains.' Partial loss of faculties, or the existence of bodily infirmities, will not suffice to invalidate a will. Thus, in one instance, certain codicils to a will were disputed on the ground that the deceased, at the time of making them, was imbecile. It was proved that the testator was liable to certain nervous attacks, and it was admitted that during these attacks he was incapable of any rational act. Moreover, he was deaf, nervous, low-spirited, and easily depressed; his eyesight was perfect, and his bodily strength was good. The only evidence, in support of his incapacity, was to the effect that his memory was defective, that he could not recognise people, and

Wills made in old age.

appeared at times to be 'lost.' On the other hand, it was shown that he was able to settle bills, to draw his own drafts, to write letters, to play cards, to go about by himself, and that he comprehended the state of his affairs. The judge in his speech said: 'Now, these accounts, with the bills regularly paid and endorsed, these drafts drawn, these counterchecks registered and marked with the date and sum for which they were drawn, the corresponding entries in the book of expenditure, prove mind and understanding, and thought, judgment, and reflection very strongly, and, in a person of his great age, of a most satisfactory and unusual degree. . . . . It is proved to my satisfaction that he possessed his mental faculties in an extraordinary degree, considering his great age, and that he had a testamentary capacity quite equal to a testamentary act of no very complicated nature.'

In another case a will was set aside, the judge deciding that, although for many purposes the testatrix might be said to have been in her right senses, she was nevertheless suffering from that failure and decrepitude of memory which prevented her having present to her mind the proper objects of her bounty, and selecting those she wished to partake of it.

When wills are made on a death-bed, it commonly happens that the medical attendant is asked to be one of the attesting witnesses. He should remember that, in acting in that capacity, he is virtually testifying that the dying person is competent to make a will, and he must be prepared to prove afterwards, if necessary, that such an opinion was correct. It is not sufficient that the dying person should assent to the draft, when read over to him; it must be certain that he understands its provisions. A dying person will often assent without fully comprehending the question put to him,

(marginal note: Wills made on the death-bed.)

but, if he can dictate the main provisions of the will correctly, there could be no dispute as to his competency to make his will.

A lunatic can give evidence in the witness-box.

Insanity does not *per se* preclude a person from giving evidence in the witness-box. This is a fact of great importance, and was clearly laid down, some years ago, on the occasion of the trial of an asylum attendant for the murder of one of the inmates. The prisoner had been convicted on the evidence of a lunatic who was an eye-witness, and the point was reserved for the higher court whether a lunatic could give evidence. In giving judgment, upholding the verdict, Lord Campbell said: 'It is for the judge to say whether the insane person has the sense of religion in his mind, and whether he understands the nature and sanction of an oath, and then the jury are to decide on the credibility and weight of his evidence. . . . . Before he is sworn, the insane person may be cross-examined, and witnesses called to prove circumstances, which might show him to be inadmissible; but, in the absence of such proof, he is *primâ facie* admissible, and the jury must attach what weight they think fit to his testimony.'

State control of lunatics.

The law recognises two methods of dealing with lunatics. A lunatic may be found such by inquisition, as the term is, and the control of his property will then be placed under the Lord Chancellor. Or he may be certified, under conditions to be alluded to immediately, to be a lunatic by two medical men, when he will come under the supervision of the Commissioners in Lunacy.

No person may receive into his house, for profit, more than one lunatic without being licensed to receive lunatics. A private lunatic may be taken care of at home by his friends, or he may be received by anyone, for profit, as a single patient, or he may be admitted into a licensed private asylum or hospital, or into licensed lodgings. When a lunatic is taken care of privately by his friends, not for profit, a Commissioner in Lunacy or other person may, upon an order from the Lord Chancellor or the Home Secretary, visit and examine the lunatic, and report upon his state.

Before a person can be received into a private asylum or hospital, or into licensed lodgings, it is necessary that certain legal formalities be complied with. An order from the nearest relative, or from some friend, must be filled up, addressed to the proprietor of such asylum, requesting him to receive the patient into his asylum, and setting forth, in the fullest manner possible, all the particulars as to the patient's name, age, abode, occupation, date of illness, previous attacks of insanity, if any, &c. This order must be signed by someone who has seen the patient within one month of the day on which it is dated, and the person signing it should state the degree of relationship (if any) between himself and the patient. *Certificate of lunacy.*

Accompanying the order must be two medical certificates. These must be signed by two registered practitioners, who must not be in partnership with each other and must have no connection with the proprietor of the asylum, and who must not be related to the person signing the order. Their examination of the patient must be conducted separately, and apart from any other practitioner, and each must state concisely the facts which he has observed indicative of insanity. They are at liberty to mention subsequently additional facts communicated to them by others, taking care to append the full name of the person supplying such information. Full marginal notes are given with the certificates, which must be strictly attended to, as any deviation from them, however trifling it may appear to be, may cause the certificate to be invalid if not amended in time. Any correction made, at the time of signing the certificate, must be initialed by the person signing. The certificates must be signed by both the medical witnesses within seven days of the patient's arrival at the asylum, though not necessarily on the same day. Within twenty-four hours of the reception of the patient, the proprietor sends a copy of the certificates up to the Commissioners in Lunacy, when, if there is any informality, the certificates will be returned for correction, which must be made within a fortnight, or the certificates become invalid, and the patient must be discharged.

In the case of a pauper lunatic, the signature of one medical man is sufficient; but the order must also be signed by a magistrate for the district before the lunatic can be received into the asylum.

Recent events have shown that not only must the letter of the law be obeyed, in regard to the signing of these certificates, but those who sign them must also act up to its spirit. This was amply shown in the various actions brought by Mrs. Weldon in 1884. In the trial of Weldon *v.* Semple, the judge left no less than seventeen questions to the jury, one of the most important being whether the examinations of the alleged lunatic had really been conducted separately, as intended by the statute. It appeared that the two doctors had gone to the house together, and had been shown in to Mrs. Weldon together. After a time one of them had retired for a little while, and when he returned the other left the room. This, it was contended, and contended successfully, was not a *bonâ fide* separate examination, as contemplated by the statute. And it was in great measure, no doubt, owing to the manner in which this examination had been conducted that the verdict in favour of the plaintiff was given.

Margin note: Weldon v. Semple.

A medical man, when asked to certify as to the lunacy of anyone, should approach such a person with a mind quite free from bias, and he must be exceedingly careful to found his inferences of lunacy on his own observation, and not on what is told him by others, of whose judgment or veracity he can seldom know much. A medical man is always liable to be sued for libel by the lunatic at some subsequent period; but if his conduct has been *bonâ fide* and straightforward he need not fear the result of any such action. An instance of the truth of this occurred in the trial of Good *v*. Whittle and others, where the plaintiff, a woman who had previously been incarcerated as a lunatic, brought an action against the defendants for libel and conspiracy. The result of the trial, however, was a verdict for the defendants on all the counts, although the certificates, which they had given, had been considered by the Commissioners as not sufficiently strong to warrant the detention of the plaintiff in the asylum. Good *v*. Whittle.

In the case of a person possessed of property there is another method of dealing with him should he be supposed to be insane. His relatives may take steps to procure an interdiction—that is, to have him declared incapable of managing his affairs. A petition to that effect must be drawn up and presented to the Lord Chancellor. This petition must be accompanied by the affidavits of at least two medical men, in which they give their opinion as to the mental condition of the person in question, and the reasons on which such opinion is based. This affidavit is more full, and requires to be more carefully drawn up, than the certificate in the ordinary lunacy form. The Lord Chancellor, if satisfied, refers the matter to one of the Masters in Lunacy, who thereupon holds an inquiry (*de lunatico inquirendo*), before a jury, if the alleged lunatic desires it. Interdiction

Commission in lunacy. The Gilbert Scott case.

The Gilbert Scott case, which was tried in the spring of 1884, is the only instance of a commission of inquiry since 1862. The necessary processes, that must be carried out, cause an enormous expenditure of money, and hence it is very rarely resorted to. In this case the petition was brought by the wife; the trial lasted several days, and created much excitement at the time. There was no doubt that the defendant had been insane, and had been confined in Bethlem, but he had escaped after a few days. The defence was that his insanity had only been temporary, and caused by drink, and that he had since recovered, as proved by his capability to conduct his business as an architect; and many witnesses were called to prove that since his release he had been carrying on his professional work with skill and success. In the end, however, the jury found that he was of unsound mind, and incapable of managing his affairs. The summing-up of Mr. Justice Denman, who tried the case, is of such importance that the following extracts are given from it:—

After reading the words of the Act of 1862, the judge said that the question for the jury was whether Mr. Scott was of unsound mind, and incapable of managing himself or his affairs. Juries had sometimes been led to the conclusion that, if a man was proved capable of doing things connected with his trade or profession, and doing them cleverly, this was sufficient to show him capable of managing his affairs. He (the judge) did not think this was so. He thought a man might be able to do all these things perfectly well, and yet be incapable of managing his affairs. When such a large word as 'affairs' was used, he should say it applied to the management of all the affairs of his life, of his wife, his children, and his home, and not to mere matters of business only. As to managing himself, it was clearly not evidence of insanity that a man was fond of drink. But if, coupled with other acts, which a jury might reasonably look upon as mad acts, there should appear an inability to control himself, on occasions when it was of the utmost importance that he should control himself, that would be evidence. In every case the presumption was in favour of sanity. A man must be taken to be sane until he was shown to be the contrary. But if a man was proved to be insane in June, and again in July, August, and September, and so on in each following month, and was then tried for murder in April, he (the judge) would never tell the jury that it was necessary still to presume sanity, as though the contrary had never been shown. If insanity had been established and had gone on for some time, then, in spite of the legal presumption, a different conclusion might be formed in the minds of those who, as reasonable men, had to judge of what followed. What was the general character of the man before all this took place? The account given of him previously might have been over-coloured. But it would be a question for them (the jury), whether the person thus depicted was not, prior to 1883, a man of singularly good appearance, manners, morals, and all such things. That being the previous character of the man, they came to the date in 1881, when he brought out his architectural work. . . . . About that time his wife noticed that he became excited and irritable, and was taking spirits. But, she said, he became all right again. (The judge then referred to the events of July 16–20, 1883, remarking that though at first it was doubtful, whether he was suffering at this time from anything more than the effects of a series of drinking bouts, as the case went on it became indubitable, that he was completely out of his mind and suffering from delusions.). . . . . There could be no doubt that down to the end of August his delusions continued. The judge then read the principal letters from Rouen, and the threatening post-cards to Mr. Dunkenfield Scott and Mr. Webster, and said this was an act which brought him within the reach of the criminal law. Would a jury, at that time, have convicted Mr. Scott of threatening to murder, or would they have found that he was, at that time, in such a state of mind that he did not understand the quality of the act he was doing? Were these sane or insane acts, and was the man, at that time, in a fit state to have the management of himself or his affairs? It was for the jury to decide these questions.

In a trial of this kind it is not necessary that the jury should be unanimous; the verdict of the majority is accepted.

The plea of insanity.   The relation of madness to crime is certainly one of the most difficult problems that come before the medical jurist. The defence of insanity is seldom raised except in cases of a charge of murder, for the very simple reason that confinement for life, in an asylum, is regarded as a greater punishment than any, that would be likely to be awarded for an offence short of murder. In the case of an idiot, or one who is clearly suffering from mania, but little difficulty will arise. But with regard to delusional insanity, or other forms than acute mania, the plea of insanity is a very difficult one to establish. By an Act passed in 1883, the jury, if they find a prisoner insane, return a special verdict to the effect that the prisoner is guilty, but insane.

Legal test.   The only test which the law regards is whether the prisoner, at the time of committing the deed, knew the difference between right and wrong. This has been objected to, as utterly insufficient, by all who have paid attention to the subject. But it

has been acted upon by the judges ever since 1843, when, after the acquittal of a man named MacNaughten for the murder of Mr. Drummond, the House of Lords submitted a series of questions to the judges on the subject of the tests for insanity. Fourteen out of the fifteen judges agreed in the replies which they gave. The following passage refers to the subject now under consideration:—
'That to establish a defence on the grounds of insanity, it must be clearly proved that, at the time of committing the act, the accused was labouring under such a defect of reason, from disease of the mind, as not to know the nature and quality of the act he was doing, or, if he did know it, that he did not know that he was doing what was wrong.'

The lunatic often knows quite well that he is doing wrong, but by some irresistible impulse he is urged on to commit the deed. The judges, however, have always been careful to explain to the jury that they must be satisfied that the prisoner did not know that he was doing wrong, and rendering himself liable to punishment, before they acquit him on the ground of insanity. No better instance could be afforded of the unsatisfactory nature of the test than that of the case of Gouldstone. In the autumn of 1883 this man drowned three of his children one afternoon, and killed the remaining two by hitting them on the head with a hammer. He then said to his wife, 'All the children are dead now. I shall be hung, and you will be single.' To the policeman he said, 'I have done it. Now I am happy and ready for the rope.' The plea of insanity was raised in his defence, and it was shown that there was a strong inheritance of insanity on his mother's side, and a remote one on his father's, and Dr. Savage gave evidence that he was of opinion, from what he had heard in court and seen of the prisoner, that he was of unsound

The case of Gouldstone.

mind, though his own short examination of the prisoner would not have enabled him to sign a certificate of his insanity. The judge told the jury 'that, as a matter of law, if the prisoner, at the time he killed the children, knew the nature and quality of the act he was committing, and knew that he was doing wrong, then he was guilty of wilful murder.' Seeing that the prisoner's own words, after the commission of the crime, showed that he knew quite well both the nature of his act, and that he would be hanged for it, the jury cannot be blamed for finding the prisoner guilty. In consequence of the strong representations which were made to him, the Home Secretary subsequently ordered Drs. Orange and Clarke to examine the convict and report to him, and in accordance with their report the prisoner was reprieved as insane, and committed to Broadmoor Asylum. Within a few months of this case there occurred a very similar one, when a man named Cole killed his child. In this case, the evidence that the man had been a dangerous lunatic for some time past was almost overwhelming, but, at the trial, the law was expounded to the jury in the same manner as in the previous case, and he was convicted and sentenced to death. The verdict was afterwards reversed by the Home Secretary, on the strength of a private official report, to the effect that the prisoner was insane.

The case of Cole.

The case of Guiteau, who murdered the late President Garfield, might further be cited as an instance in which the plea of insanity was rejected, on the ground that the prisoner knew he was doing wrong, and would be punished for it. Had the victim been less distinguished, it is possible that in this case also the obvious imbecility of the convict would have secured for him an acquittal on the ground of irresponsibility.

The case of Guiteau.

Two particularly dangerous forms of mania, as regards the tendency to homicide, are epileptic mania and puerperal mania. The latter is the more generally recognised by the public, but the former is the more dangerous, inasmuch as the attacks are sometimes exceedingly sudden. Dr. Savage, in his recent work on insanity, when speaking of epilepsy, says that it is common to meet with cases in which, immediately before or immediately after a fit, an outburst of uncontrollable fury, of the most destructive kind, takes place. These cases he considers to be, without exception, the most dangerous of all lunatics. A good instance was afforded of this form of homicidal mania in the case of a young man, who murdered his sweetheart at Woolwich, in December 1884. Whilst in prison awaiting his trial he had several epileptic fits, in consequence of which he was examined, upon the order of the Home Secretary, by two experts in lunacy, who reported that he was of unsound mind. When the time for his trial arrived, the grand jury found a true bill against him, but, upon the order of the Home Secretary, he was transferred to a lunatic asylum, and was not called upon to plead. *Puerperal mania. Epileptic mania.*

The possibility that the prisoner is feigning insanity must always be borne in mind. It is worth noting, as Taylor justly remarks, that people do not feign insanity to avoid suspicion, and consequently, in a pretended case, the insanity does not make its appearance until after the charge is brought against the prisoner. Inherited tendency to insanity, though undoubtedly affording an increased probability of the reality of the insanity, should not be allowed to have too much weight. The mode of onset, and previous mental state of the alleged lunatic, should be carefully investigated, and he should be most closely watched. The real lunatic is sure to be consistent throughout, the *Feigned insanity.*

would-be lunatic is sure not to be. A real lunatic generally strenuously denies his insanity when this is pleaded for him; he who is only pretending to be insane would never risk doing this.

The legal relationships of intoxication, Intoxication, when voluntary, is no justification to a criminal offence, though it may be perfectly true that the perpetrator of the crime has no recollection of the occurrence. A person, therefore, who commits a murder whilst in a state of intoxication, cannot plead this fact in extenuation of his crime, excepting, perhaps, where a weapon was used which would not be likely to cause death, and where no malice or premeditation existed. Such circumstances *might* reduce the crime from murder to manslaughter. Contracts or wills made by a person in a state of intoxication, would be voidable by the party when in his sober senses, but would be binding unless repudiated by him without delay. If, however, the intoxication was not voluntary, but was induced by others with a view to obtaining consent to a contract, then such contract would be treated as having been obtained by undue influence and fraud, and would be invalid. The same ruling would apply in the case of a will.

and of delirium tremens. A person is not responsible for his actions whilst in a state of delirium tremens, even although this is always the result of habits of intemperance, which must be assumed to have been voluntarily induced.

CHAPTER XVIII.

POISONS.

The legal relationships of poisons may be gathered from the following statutes of the Criminal Consolidation Act: 'Whosoever shall unlawfully and maliciously administer to, or cause to be administered to or taken by, any other person, any poison or other destructive or noxious thing, so as thereby to endanger the life of such person, or so as thereby to inflict upon such person any grievous bodily harm, shall be guilty of felony,' and 'Whosoever shall unlawfuly apply or administer to, or cause to be taken by, or attempt to apply or administer to, or attempt to cause to be administered to or taken by, any person, any chloroform, laudanum, or other stupefying or overpowering drug, matter, or thing, with intent, in any of such cases, thereby to enable himself or any other person to commit, or with intent &c. to assist any other person in committing any indictable offence, shall be guilty of felony.' The term 'destructive or noxious thing' would obviously include many substances which, from a medical point of view, are not poisonous at all—such, for instance, as powdered glass, molten lead, boiling water, none of which can be considered as having any poisonous properties, the ill effects due to their administration being

Statutes relating to poisons.

caused by a purely mechanical action. But the law looks not so much to the nature of the substance, as to the intent with which it was used.

Medical definition. Many medical definitions have been proposed. Thus Taylor defines a poison to be a 'substance which, when absorbed into the blood, is capable of seriously affecting health or of destroying life.' Guy says a poison 'is any substance or matter (solid, liquid, or gaseous) which, when applied to the body outwardly, or in any way introduced into it, without acting mechanically, but by its own inherent qualities, can destroy life.' Winter Blyth, the latest writer on the subject, says, 'A substance of definite chemical composition, whether mineral or organic, may be called a poison, if it is capable of being taken into any living organism, and causes, by its own inherent chemical nature, impairment or destruction of function.'

Mode of action. This leads to the consideration of the mode of action of poisons, and, at the same time, indicates the method by which this action takes place, viz. by absorption. That poisons are absorbed into the blood, and thus reach the various parts of the body through the circulation, was proved many years ago by means of experiments. When poison was put into the stomach of an animal, it was found, that no symptoms were produced if the vessels leading to the liver had been previously ligatured, but that if such ligature was removed the poison began to act within a minute. Another experiment was the removing of a limb, and leaving it connected with the body only by means of quills inserted into the blood-vessels. If the limb was then poisoned, the poison acted on the whole body, and must have reached it by means of the circulation; whereas when the limb was connected with the trunk by the nerves only, and not by the blood-vessels, the introduction of poison into the limb was not followed by any poisonous effects on the body generally.

Poisons are most commonly introduced into the body by the mouth, but they may also enter by the respiratory passages, by the rectum, vagina, eye, ear, skin or subcutaneous tissues. As a general rule, it may be said that poisons act more readily when introduced into the stomach than through the unbroken skin, and most readily of all when injected into a vein. The causes which modify the action of poisons may be considered under two heads, viz. causes connected with the poison itself, and causes dependent on the person subjected to its influence. *[Causes modifying the action of poisons.]*

1. Causes connected with the poison itself. Under this head it must be noted that the quantity employed exerts, as might be expected, a very important influence on the effects, some poisons in large doses acting not only more speedily, but producing their results through different channels. Again, poisons are far more active in the finely-divided state; thus their effects are much more marked when they are given in a state of solution. The activity of some poisons is much increased by their entering into chemical combination with other substances, whilst other poisons, on the other hand, lose all their dangerous characters by such a union. Admixture with other substances, if it renders the poison more soluble, makes it more active. It might be supposed that dilution would diminish the activity, but this is not always the case, for dilution enables the poison to be more readily absorbed, and thus, as in the case of oxalic acid, facilitates its action on the heart and nervous system.

2. Causes connected with the person. *Idiosyncrasy* usually takes the form of undue susceptibility to the influence of certain drugs. Alcohol, *[Idiosyncrasy.]*

mercury, opium and tobacco will occur to everyone as being those in respect of which idiosyncrasy is most often exhibited. Certain articles of food, too, prove poisonous to some, though eaten by the majority with impunity. Idiosyncrasy may, however, betray itself by exactly the opposite condition— viz. an unwonted insusceptibility to the action of substances usually regarded as poisonous. The tendency of *habit* is to diminish the effects of the poison, and enable a person to take a relatively large dose without danger. Probably there is no poison to which the system may not be gradually accustomed, though it is open to doubt whether tolerance of any poisonous drug can be obtained in this way, without some organ or tissue in the body ultimately paying the penalty. Tobacco and opium, amongst the organic poisons, naturally occur to the mind as drugs, tolerance of which is attained without much difficulty ; whilst, with reference to inorganic substances, mention may be made of the old man at Constantinople, who was in the habit, for thirty years, of swallowing enormous doses of corrosive sublimate, until his daily allowance reached a drachm, and who, nevertheless, lived to be a hundred years old. The effect of certain *diseased states of the body* is to impair the activity of a poison, or rather to render the body less susceptible of it. Thus, in dysentery and cholera, opium may be given in doses that would be highly dangerous to a person in health. In tetanus and hydrophobia, too, the action of drugs seems to be much weakened. On the other hand, there are certain diseases which render the administration of particular drugs a matter for more than ordinary caution; in Bright's disease, for instance, where the function of the kidneys is more or less impaired, the use of mercury or opium is attended with considerable risk.

Margin notes: Habit. Disease.

Various schemes have been proposed for the classification of the poisons. Some writers group them according to their actions, either physiological or chemical; others have proposed a mere arbitrary division according to the three kingdoms from which the poisons are derived. Taylor has only two groups, viz. irritants and neurotics; whilst Christison makes of this latter group two classes, narcotics and narcotico-acrids. It is not easy to devise a classification which would meet all the requirements, and Taylor's, being the simplest, will be followed. *Classification of poisons.*

The mode of action of the various poisons is, in many cases, so different that it is not possible to lay down any points of general evidence of poisoning, that would serve in all cases. Taylor thus sums up the steps to be taken, in the case of suspected poisoning: 1. The exact time at which the symptoms occurred, as well as their nature, should be ascertained. 2. The exact period at which they were observed to take place after a meal, or after food or medicine had been taken. 3. The order of their occurrence. 4. Whether there was any remission or intermission in their progress; or whether they continued to become more and more aggravated until death. 5. Whether the patient had laboured under any previous illness. 6. Whether the symptoms were observed to recur more violently after a particular meal, or after any particular kind of medicine. 7. If vomiting had taken place, the vomited matters, especially those first ejected, should be procured; their odour, colour, and acid or alkaline reaction noted, as well as their quantity. 8. If none were procurable, and the vomiting had taken place on an article of clothing, or furniture, or on the floor, then a portion of the clothing, sheet, carpet, &c. might be cut out and preserved for analysis; if the vomiting had occurred *General evidence of poisoning.*

on a deal floor, a portion of the wood might be scraped or cut out; if on a stone pavement, then a clean sponge soaked in distilled water might be used to secure any traces of the substance. The vessels in which vomited matters had been contained would often furnish valuable evidence, since heavy mineral poisons may adhere to the bottom or sides. 9. The nature of the food or medicine last taken should be ascertained, as well as the exact time at which it was taken. 10. The exact nature of all the different articles of food used at the meal should be ascertained. 11. Any suspected articles of food, as well as the vomited matters, should be sealed up as soon as possible in clean glass vessels, labelled and reserved for analysis. 12. Any explanation voluntarily offered by those present, or by any who might have been concerned in the supposed poisoning, should be noted down in the identical words used. 13. Inquiry should be made as to whether others partook of the food or medicine, and whether they showed any symptoms of being affected thereby. 14. Inquiry should also be directed to determine whether the same food or medicine had been taken on other occasions by the patient or others giving rise to ill effects.

Evidence from physiological experiments.
In addition to the evidence supplied by the symptoms and examination of the ejecta, proof of poisoning may be obtained from the post-mortem appearances, from chemical analysis of the viscera, and from physiological experiments performed with the extracts made from the ejecta or viscera of the dead person. This latter class of evidence is absolutely necessary in the case of the alkaloids, for which no chemical tests are known; and the conviction of Lamson, in 1882, for poisoning his brother-in-law with aconitine, could not have been obtained had this method of investigation not been adopted.

Cases of imputed poisoning are, fortunately, not common, nor are those of pretended poisoning of more frequent occurrence; in either case the investigation would have to be a searching one before the true nature of the case could be determined. *Imputed and pretended poisoning.*

Before proceeding to deal with the poisons in detail, a few general remarks on treatment will not be out of place. The first thing for a medical man to do, when called to a person who has just taken poison, is to get the poison out of the stomach as quickly as possible. With this object an emetic would be given, and the stomach washed out with the stomach-pump; the contra-indications, to this method of treatment, will be mentioned under the head of the different poisons. The next thing is to neutralise what poison has not been removed in this way: this is effected by means of chemical, mechanical, and physiological antidotes. The poison may, further, be got rid of by elimination through the intestines, kidneys or skin. Any unusual symptoms, or complications that arose, would have to be treated according to general principles. *General remarks on treatment.*

CHAPTER XIX.

IRRITANT POISONS.

THE MINERAL ACIDS.

Poisons.   UNDER the head of Irritant Poisons come the mineral acids, the alkalies, the alkaline earths, the metals, and vegetable and animal irritants. The mineral acids are sulphuric, nitric and hydrochloric acids. The strong acids all have a corrosive action, and do not give rise to specific remote effects. It will be seen that their mode of action, and the appearances to which they give rise, are very similar.

Sulphuric Acid.

Sulphuric acid, or oil of vitriol, is a heavy oily-looking liquid, colourless when pure, or of a brownish colour when impure; it does not fume when exposed to the air, and when mixed with its own bulk of water, great heat is given out. Poisoning by sulphuric acid is not often homicidal, owing to the difficulty in administering it; it is occasionally taken by accident, especially by children, and, less often, resorted to by suicides.

Symptoms.   The symptoms will depend upon whether the strong acid has been taken, and upon the amount swallowed, as well as upon the condition of the stomach at the time of taking it as regards food. This will exercise a most decided influence upon the symptoms and result. When the strong acid is taken the symptoms come on *at once*, whilst the acid is being swallowed. At first a sour taste is perceived, which is immediately followed by a burning pain in the mouth, throat and gullet, extending to the pit of the stomach; the patient is seized with eructations and vomiting of a frothy blackish matter, consisting of mucus, blood, and flakes of epithelium; the pain may be so intense that the patient speedily becomes collapsed. The lips are shrivelled and brownish, occasionally they appear as if blistered, the mucous membrane of the inside of the mouth and of the tongue is in a somewhat similar state, being whitish and more or less shreddy; the teeth become loose and discoloured. Sometimes the tongue appears as if it was smeared with white paint. There will also, probably, be stains about the mouth and neck. The breathing is often laboured, partly owing to affection of the larynx by the acid, and partly to spreading of the inflammation from the stomach to the neighbourhood of the diaphragm; and there may be aphonia and stridor. Vomiting, as a rule, sets in immediately, and may be incessant. The matters first vomited usually contain the poison, and if they fall on any article of clothing, they may, by their corrosive action, suggest the nature of the poison used. The pulse is throughout weak, and generally, but not necessarily, increased in frequency. The skin is clammy, and the countenance pale and anxious. There is great dysphagia, with intense thirst, obstinate constipation, tenesmus and dysuria. Sometimes there are convulsive movements of the muscles, especially those about the mouth. As a rule, the intellect remains clear to the last, the patient dying of collapse, more or less suddenly, in about twenty-four hours. Occasionally there may be delirium or convulsions. In one case, where death took place quite suddenly, this was found to be due to the separation of the mucous membrane of the stomach, which had taken place in one mass. The urine will be found to contain a great excess of sulphates, some albumen and hyaline casts, and is always scanty in quantity.

If, however, the dose taken has not been such as to produce so rapidly fatal a result, there will supervene fever, with a dry skin and a frequent pulse. The vomiting will continue, and the vomit will perhaps contain altered blood and shreds of mucous membrane. The breath becomes fœtid, and salivation is often a prominent feature of the case. The abdomen is tense, tender, and distended, as it may be also in the more acute form, and the patient suffers from cramps in his limbs. There is dysphagia varying in degree, impaired digestion, and consequently emaciation. The patient may die, after some weeks or months, of gradual starvation, or he may recover to be a confirmed dyspeptic for the rest of his life, or recovery may be complete. Death supervenes in about twenty-four hours if the poison is going to prove fatal speedily, whilst a person has been known to die from its effects after eleven months. The most speedily fatal case on record is that of a man, aged fifty, who died three-quarters of an hour after swallowing three ounces and a half of the strong acid. One drachm killed an adult in a week, and forty drops proved fatal to a child in twenty-four hours.

Fatal period and fatal dose.

Post-mortem appearances.

About the angles of the mouth, on the chin and neck, there will, probably, be some yellowish or brownish stains, and there might be similar stains on the fingers. The internal appearances will depend in great measure upon the quantity taken, and the time at which death occurred. If the case had proved rapidly fatal, the mucous membrane of the mouth, tongue, and pharynx would be found white, swollen, sodden, softened, and in places corroded; there might be swelling of the epiglottis, and of the mucous membrane of the whole larynx. The mucous membrane of the œsophagus is generally in a similar state to that of the mouth; ulceration may be present, but perforation is very rare. The stomach may be distended with gas, or it may be empty and contracted, or it may be empty and collapsed from perforation. The contents, on opening the stomach, will be found to be strongly acid if no antidote has been given; they consist chiefly of a glutinous fluid, yellowish brown or blackish in colour, probably composed of mucus and altered blood. The mucous membrane of the stomach is blackened, and often thrown into rugæ; the black colour is in the membrane itself, and cannot be removed by washing. When perforation has taken place, some of the contents of the stomach should be found in the abdominal cavity; often the perforation takes place during the necessary manipulation in removing it. The holes in true perforation are more or less round, and the coats in the immediate neighbourhood are thinned, discoloured, and disintegrated. The duodenum often suffers almost to the same extent as the stomach, and the rest of the intestines will be sometimes found to be corroded. There is often extreme redness of the peritonæum, and sometimes effusion of lymph on its surface. It should be remembered that the corrosive action of the acid would take place just as readily after death as during life, and therefore the changes, found at the time of the post-mortem examination, must not be assumed to represent the exact state of affairs at the moment of death. Putrefaction is delayed. In one case—that of a child—the mouth was unaffected. This was attributed, no doubt correctly, to the poison having been poured into the child's throat with a spoon.

o

Appearances in the later stages.

When the patient has survived the immediate effects of the acid, the appearances will depend upon the length of time he has survived. In a case where a man died two months after taking the poison, the stomach was found small and contracted, with many adhesions to the pancreas and liver; there were several transverse rugæ, the walls were thickened at the smaller curvature, the inner surface was covered with a layer of pus, with no trace of mucous membrane, and was everywhere pale red, uneven, and crossed by cicatricial bands. In two parts, at the greater curvature, the mucous surface was strongly injected in a ringlike form, and in the middle of the ring was a funnel-shaped ulcer. The gullet was contracted at the upper part, and then gradually widened out.

Treatment.

If the stomach is empty at the time the poison is swallowed, there will be very little chance of saving the life of the patient, as the corrosive action will commence at once. The best chemical antidote is carbonate of magnesia, but all the alkaline bicarbonates are useful, mixed with water or milk and water. Whilst this is being prepared, milk, or milk and water, oil, soap and water, may be given, or gruel. It is important to neutralise the action of the acid by dilution, as speedily as possible. Dinneford's fluid magnesia or chalk may also be used. If none of these are immediately available, a little plaster should be scraped off the ceiling, and mixed with water into a paste, and given to the patient. Although it is very desirable that the patient should vomit, emetics must not be given, nor must the stomach-pump be used on any account. In many cases there will be very great difficulty in swallowing; in such it seems justifiable that an attempt might be made to introduce the antidotes and diluents into the stomach, by means of a soft catheter passed into the nose; no

SULPHURIC ACID.

attempt, however, should be made to pass this further than into the upper part of the œsophagus. Should the patient survive, the free use of alkaline diluents should be persevered with for some days, and no solid food should be allowed for several days. The use of opiates, of leeches to the pit of the stomach, or of hot fomentations, would have to be determined upon general principles. Tracheotomy might be called for, if the laryngeal symptoms were very threatening.

Strong sulphuric acid is easily recognised. It chars or blackens wood, the spot remaining moist. When some sulphuric acid is added to wood scrapings, copper filings, or mercury, in a test-tube, sulphurous acid is given off, which may be readily recognised by the pungency of its fumes, as well as by its property of liberating iodine, and its bleaching properties. These are shown by dipping some starch paper into a solution of iodic acid, and exposing it for a moment to the fumes, when the paper is turned blue, the free iodine combining with the starch to form the iodide of starch; if the paper be further exposed to the fumes, the blue colour disappears, and the paper is bleached white. Dropped on paper, sulphuric acid makes a black stain, and ultimately chars the paper; the dilute acid will also produce this effect if, after it has been dropped on the paper, the latter be held close to the fire. The dilute acid may be recognised by the following test: With nitrate of barium it gives a white precipitate (sulphate of barium), insoluble in nitric acid. The precipitate should be collected and heated in the flame of a blowpipe; the sulphate is thus reduced to a sulphide; this is then placed on a piece of moistened lead-paper, and touched with dilute hydrochloric acid. Sulphuretted hydrogen is thus formed, and thence sulphide of lead, blackening the paper. Linen is

Tests.

charred and corroded by the strong acid; woollen garments are rendered damp and rotten; cloth is destroyed by it, the spot being of a dull red colour and moist. Indeed, this dampness of the spot is the chief characteristic of a burn due to sulphuric acid. To examine a suspected spot, it should be cut out and soaked in water. The solution thus obtained is strongly acid, and gives the reactions of dilute sulphuric acid. Dropped on iron, sulphuric acid speedily attacks it, forming a sulphate which may be dissolved out by water.

The examination of organic compounds for sulphuric acid, free or combined, is not a process that can be undertaken, except by one skilled in chemical manipulations. Full details of the various steps will be found in the text-books on poisons.

Nitric Acid.

Nitric acid is known in the commercial world as Aqua Fortis. When pure, it is almost colourless, but very often it has a yellow or yellowish-red tint, whence its name of 'red spirit of nitre.' Poisoning seldom occurs from nitric acid, except as the result of suicide or accident.

Symptoms. The symptoms are practically the same as those produced by sulphuric acid. Gas is always developed in the alimentary canal, the stains about the lips and hands are more yellow, the teeth are often affected, being rendered very white, and the enamel is sometimes attacked by the acid. About half the cases prove fatal. The fumes of nitric and nitrous acids are very irritating, and have, on a few occasions, proved fatal to persons inhaling them. Thus, in one case, a chemist was pouring a mixture of nitric and sulphuric acids from a carboy, when this broke, and he was exposed, for a few minutes, to the fumes. For three hours he seemed to suffer no ill effects from this, but was then seized with cough, difficulty of breathing, and tightness of the chest; these symptoms increased, and he died eleven hours after the accident. The trachea and bronchi were much congested, the latter containing some blood; the lining membrane of the heart and aorta was slightly inflamed.

The stains about the lips will be more yellow than in a case of sulphuric acid poisoning. There may be a yellow frothy liquid about the mouth and nostrils, and the mucous membrane of the mouth, œsophagus, and stomach may have a yellowish tinge. The larynx is more often affected than in the case of sulphuric acid, and perforation of the stomach is less common. *[Post-mortem appearances.]*

The treatment does not materially differ from that recommended in the case of sulphuric acid poisoning.

The concentrated nitric acid fumes when exposed to the air, on organic substances it leaves a yellow stain. When mixed with copper filings, a red vapour of nitrous acid is given off, and a greenish-coloured solution of nitrate of copper formed; it does not dissolve gold leaf even on boiling, but, on the addition of a little hydrochloric acid, the gold is immediately dissolved. Dilute nitric acid may be known from sulphuric or hydrochloric acids, by its not giving a precipitate with salts of barium or silver. When not in the free state it is usually present as a nitrate, and may be recognised by the following tests: To a nitrate dissolved in water, a crystal of ferrous sulphate is added and partially dissolved; some strong sulphuric acid is then poured very gently to the bottom of the test-tube; three layers should appear, the upper one being the nitrate solution, the middle one the ferrous sulphate, and the lowest the sulphuric acid; if a nitrate is present, the middle layer *[Tests.]*

should be of a greenish-black hue. In examining the contents of the stomach, if these have not been already neutralised, some bicarbonate of potash should be added, and the mixture boiled and filtered; the residue may be evaporated to dryness, and treated with strong alcohol; the filtrate thus obtained may be tested, in the ordinary way, for nitric acid. On animal fabrics the stains appear yellow, and red upon articles dyed with vegetable colours. The stained cloth must be boiled in a little water, and the solution tested as above. A yellow stain might also be produced by bile or iodine; to distinguish between these, it will be sufficient to add a weak solution of caustic potash; the nitric acid stain becomes orange, the bile stain is unaltered, and the iodine stain disappears.

Hydrochloric Acid.

This is also called muriatic acid, and is known as the spirit of salt in commerce. Poisoning by this substance is rare, as compared with the other acids, and is generally the result of accident.

Symptoms. The symptoms are much the same as in poisoning by the other acids, but hydrochloric acid does not stain the skin in the same definite manner, and, being very volatile, is more likely to set up laryngeal symptoms. The following is one of the most rapidly fatal cases, that has occurred in this country: A child, aged two years, swallowed some spirit of salt of commerce, the quantity being roughly guessed at half a teaspoonful; he immediately commenced crying, and shortly after was sick, the vomited matters being somewhat black. The extremities were cold and the pulse feeble, there was no cough, or evidence that any of the poison had entered the larynx, the lips were not

HYDROCHLORIC ACID.

shrivelled or blistered, but there were one or two small excoriations about the angles of the mouth. The tongue was somewhat shrivelled on its surface, and white. Swallowing was evidently painful; he vomited the antidotes that were given, and the vomiting continued until his death, which took place, from collapse, about four hours after taking the poison. The speedy death, in this case, was probably due to the tender age of the patient. Death has been known to occur in two hours. Recovery is not common.

The changes found after death do not differ, in any important respect, from those of poisoning by the other mineral acids, and the treatment would be that already indicated for sulphuric acid.

When hydrochloric acid is exposed to the vapour of ammonia, dense white clouds are given off. When boiled with peroxide of manganese, chlorine is given off, and may be known by its bleaching properties. With nitrate of silver, hydrochloric acid gives a dense white precipitate, insoluble in nitric acid, but soluble in ammonia. This precipitate turns purple or blackish in the presence of the light. When heated, it fuses into a soft yellow liquid, which on cooling becomes a horny mass. The only salt it could be confounded with would be the cyanide of silver, which, however, is soluble on the addition of nitric acid in excess, and which gives off cyanogen gas on heating, as shown by its inflammability. The urine should always be examined for chlorides; (10 per cent. of the solid extract is the normal ratio in healthy urine.) It must not be forgotten that hydrochloric acid exists normally in the gastric juice; in health, probably there is none in the free state, but, in certain forms of dyspepsia, an appreciable quantity may be present. If, however, the quantity of acid discovered on analysis was other than small, it could not all

Tests.

be accounted for in this way. On black cloth hydrochloric acid produces a green stain, on other articles of clothing a red one, the cloth is never corroded, and the stain lacks the moisture so characteristic of the effects of sulphuric acid. If the stained part be cut out and boiled, the solution could be tested for the acid.

CHAPTER XX.

THE ALKALIES AND ALKALINE EARTH, OXALIC ACID.

Caustic Potash and Soda.

THE effects of caustic potash and caustic soda, of carbonate of potash and carbonate of soda, are so similar, that one general description will serve for all. These substances do not act as corrosives, except when in a very concentrated state, and they do not give rise to specific remote effects.

At first there is an unpleasant taste, followed by a burning pain extending from the mouth to the stomach, with great tenderness at the epigastrium. Vomiting is generally present, the vomit consisting of tenacious blood-stained mucus, and of a brown grumous matter with flakes of epithelium; it is always strongly alkaline, and has a soapy, frothy appearance. Purging may set in with severe colic. Sometimes there is collapse, which may be followed by death. If the patient survives this stage, as he commonly does, there may be inflammation of the mouth and gullet, followed by dysphagia, stricture of the œsophagus, and death from starvation. The most rapidly fatal case known is that of a boy who died three hours after swallowing three ounces of a strong solution of carbonate of potash. Much would depend upon the emptiness or otherwise of the stomach, as to the dose that would kill; forty grains

Symptoms.

have been known to prove fatal. Probably half an ounce may be regarded as a poisonous dose under ordinary circumstances.

Post-mortem appearances. When death has taken place in a few days, the mucous membrane of the mouth, throat, gullet and stomach will be found whitish, softened and corroded; sometimes ecchymoses are present. It is doubtful whether perforation ever occurs. In cases where death has taken place after a long time from gradual starvation, the appearances are those of cicatrisation and contraction of the parts originally damaged. Thus, in one case the throat and upper part of the gullet presented nothing abnormal, but the gullet became thicker, and more contracted, as it descended to the stomach, so that the opening into the stomach would scarcely admit a crow-quill, owing to thickening of the mucous membrane. In all cases, whether recent or chronic, the larynx should be examined, as it is very liable to be damaged.

Treatment. Vinegar and water or lemon-juice should be given without delay to neutralise the poison; olive oil in large doses does good by converting the alkali into a soap. Oranges, milk, gruel, and barley-water may be freely used. Opium or stimulants would be administered according to the indications.

Tests. Alkalies may be distinguished from other substances, by their strongly alkaline reaction, by their solubility in water, and by the fact that they are not precipitated by sulphide of ammonium or carbonate of ammonium. The caustic alkalies may be known from their carbonates by the latter effervescing on the addition of dilute hydrochloric acid, and also by the addition of nitrate of silver, the caustic alkalies giving a brown precipitate, their carbonates, a yellowish one. Potash salts may be known from sodium salts by giving a violet flame in the Bunsen burner, and not a yellow one. Salts of potash are precipitated by tartaric acid, and with platinic chloride they give a precipitate insoluble in strong alcohol, whilst that formed with sodium is readily soluble in alcohol or water. The contents of the stomach are usually frothy, possess a soapy feel, and a strongly alkaline reaction. They should be evaporated to dryness, and then heated, to char the animal and vegetable matter; the incinerated residue should then be digested with alcohol; the solution will contain the alkali in the form of a carbonate.

Ammonia.

Ammonia exists in the form of a gas, or it may be met with as a solution of this gas in water. Under a pressure of six-and-a-half atmospheres, at the ordinary temperature, it is readily liquefied.

After swallowing ammonia, or exposure to its vapour, the symptoms will be irritation, redness, and swelling of the mouth, fauces and tongue, pain extending down the gullet to the stomach, and vomiting; salivation is usually produced. In addition, there are often present dyspnœa, hoarseness, and cyanosis, which often lead to a fatal result. If the patient survives the immediate effects, stricture of the œsophagus may result. The most rapidly fatal case on record is that of a man who, having been bitten by a mad dog, took a mouthful of spirit of ammonia and died in four minutes. *[Symptoms.]*

The following were the changes found in the body of a man, who died on the third day after swallowing a strong solution of ammonia. The mucous membrane of the tongue was softened and had peeled off, the lining membrane of the trachea and bronchi was softened, and covered with layers of false membrane, the larger bronchi were com- *[Post-mortem appearances.]*

pletely obstructed by casts of this membrane. The mucous membrane of the œsophagus was softened, and at the lower end was completely destroyed. There was a hole in the anterior wall of the stomach, through which its contents had escaped; the edges of the hole were soft, ragged, and blackened. The vessels were injected with dark-coloured blood. No trace of the poison was found anywhere. Sometimes there is œdema of the glottis.

Treatment. Dilute solutions of acetic acid or lemon-juice should be given with demulcents, such as milk or barley-water. The rest of the treatment should be symptomatic.

Tests. If a rod dipped in hydrochloric acid be exposed to the vapour of ammonia, a white deposit forms on the rod. Nessler's test is the most delicate; it is performed as follows: To some perchloride of mercury in solution, iodide of potassium in excess is added, and then free potash in excess; the merest trace of ammonia will give this solution a brown discoloration. Ammonia is separated from organic liquids by distillation; if the liquid is not alkaline previous to distilling, it should be rendered so by the addition of burnt magnesia. If the distillate gives a brown colour with Nessler's test, and if, after the solution thus obtained has been neutralised by sulphuric acid and evaporated to dryness, a crystalline mass is left, which gives a copious precipitate with platinic chloride, hardly at all soluble in alcohol, ammonia is present.

Carbonate of ammonia, linimentum ammoniæ, linimentum camphoræ compositum, all give rise to the same symptoms as pure ammonia when taken internally in any quantity, and do not call for any separate notice.

Nitrates of Potassium and Sodium.

Nitrate of potash, or nitre, as it is commonly called, is well known, and has several times proved fatal, through being taken in mistake for Epsom salts. Sodium nitrate has not proved fatal to man, but it acts in the same way, and the description of the effects of one is applicable to the other. After a moderately large dose, there may be a sense of uneasiness in the abdomen, with vomiting and diarrhœa, and frequent micturition. If the dose has been larger, there may be great pain in the abdomen, with violent vomiting and purging, the vomited matters and fæces being stained with blood. The pulse becomes irregular and very slow; there are cold sweats; the patient has cramps, especially in the calves of his legs, convulsions, falls into a state of collapse, and dies. The effects are, however, somewhat uncertain, as recovery, even after two ounces of the salt have been taken, has been known. The nitrates are supposed to act partly by being converted into nitrites, and entering the circulation as such; and this is supported by the fact that unpleasant, and even alarming, symptoms have been known to result from the medicinal use of nitrite of sodium. *Symptoms.*

On post-mortem examination, the mucous membrane of the stomach, duodenum, and upper part of the small intestines is inflamed, discoloured, and sometimes softened. The contents of the stomach are often mixed with blood. The mucous membrane of the stomach may be extensively detached, and perforation has been recorded. *Post-mortem appearances.*

As regards treatment, vomiting should be encouraged as much as possible, but great caution would be necessary in the use of the stomach-pump. *Treatment.*

Tests.
There is no chemical antidote; mucilaginous drinks should be given, and opiates if necessary. When heated in a test-tube with some copper filings, sulphuric acid, and water, nitre evolves red fumes of nitric oxide.

Chlorate of Potash.

Symptoms.
When administered in too large a dose, this has an irritant and also a depressing effect, and several deaths have occurred from its medicinal use. In four children, who died from the effects of taking a large quantity at very short intervals, there were vomiting and diarrhœa in some, but collapse, with lividity and feeble pulse, was the most prominent symptom. In adults, death has usually resulted from nephritis, after the lapse of a few days. The smallest dose, that has been known to prove fatal, is 46 grains, which killed a child of three years. The most rapidly fatal case was that of a woman, who died five hours after taking the poison. It would probably prove more readily dangerous to anyone who had diseased kidneys, than to a person in health. Chlorate of potash causes a peculiar change in the blood, destroying the hæmoglobin and dissolving the corpuscles, so that they cease to act as oxygen carriers. In only one of the children, above referred to, was there an autopsy made, when a large ecchymosis was found on the mucous membrane of the stomach, as if it had been burnt with an acid; the spleen was engorged and friable; the kidneys were swollen.

Post-mortem appearances.

Treatment.
The treatment would depend partly upon the time that had elapsed since the poison was taken, and also upon whether the irritant or depressant symptoms predominated, and would be directed accordingly.

Tests.
The solution to be tested is acidulated with a

few drops of sulphuric acid, and then some sulphate of indigo is added, till the solution has a blue colour; a few drops of sulphurous acid are then added, when, if any chlorate of potash or soda be present, the blue colour immediately disappears.

Sulphate of Potash.

When taken in large doses, this gives rise to the symptoms of an irritant poison. For a long time it was supposed to be an almost inert drug, but a few deaths after its administration have occurred. In the fatal cases it has mostly been used for the purpose of procuring abortion, and most of the cases have been in France.

Severe pain in the stomach, nausea, vomiting, purging, cramps in the limbs, and collapse have been amongst the more prominent symptoms. In one case death is said to have occurred in an hour and a half, and in another in two hours. The number of recorded cases is not sufficiently great to enable an opinion to be formed as to the fatal dose. *Symptoms.*

In one instance, the stomach and upper portion of the small intestines were of a deep purple colour. The stomach, when opened, showed marks of irritation, and its mucous membrane was much congested. The treatment should consist in encouraging vomiting, by means of emetics, and in the use of the stomach-pump; common salt or chloride of lime might be given internally. *Post-mortem appearances.* *Treatment.*

In solution it is precipitated by chloride of platinum, the precipitate being of a canary-yellow colour; and also by tartaric acid, in the form of granular crystals. The sulphate could be recognised by the addition of a barium salt. *Tests.*

Alum.

This is a sulphate of alumina and potash. When swallowed in large quantities, it may give rise to the symptoms of an irritant poison. Only one fatal case has been recorded from its effects. A young man, aged 27, swallowed an ounce; the symptoms were: a burning sensation and a feeling of constriction at the pit of the stomach, nausea, and vomiting. There was no diarrhœa, the intellect remained clear, and death occurred in eight hours. At the autopsy, the whole of the lining membrane of the digestive tract was found to be inflamed. Emetics should be given and the stomach-pump applied without delay. As antidotes, hydrate of magnesia and a dilute solution of carbonate of ammonia have been recommended. A solution containing alum is acid, and is unaffected by ferrocyanide of potassium or sulphuretted hydrogen. The sulphuric acid gives a white precipitate with nitrate of barium; the alumina gives a gelatinous precipitate with ammonia, and the potash gives a crystalline one with perchloride of platinum.

Salts of Barium.

Symptoms. The chloride and the carbonate are the most important, the sulphate being highly insoluble, and therefore unlikely to act as a poison. Poisoning by these is mostly due to accident. Nausea and vomiting, with pains in the stomach, are among the earliest symptoms; then muscular weakness, noises in the ears, disturbances of vision, cramps, and even paralysis. Out of fifteen recorded cases, death took place in nine. A drachm of the carbonate proved fatal in one instance. The stomach and intestines will be found more or less inflamed; the brain is sometimes congested. In one case, perforation of the stomach was found, but it is doubtful whether this was to be attributed entirely to the effects of the barium. The treatment would be the same as for the other irritant poisons. Soluble sulphates, such as the sulphate of magnesia, would constitute the best antidote. Barium salts may be recognised by the green flame they give when heated in a Bunsen burner. Sulphate of barium is exceedingly insoluble. If a substance is to be examined for barium, it should be treated with boiling water, and then concentrated; sulphuric acid should then be added, and the precipitate collected. If this precipitate should be found to contain sulphate of barium, it is clear that this must have been obtained from some soluble salt, such as the nitrate or chloride. Next, some of the substance, left after the treatment with water, should be treated with hydrochloric acid, and filtered; if, after the addition of sulphuric acid to the filtrate, sulphate of barium be found, then it must have been present in the form of a carbonate. Lastly, the inorganic mass may be burnt, and the ash fused with carbonate of soda; if this be dissolved in hydrochloric acid, and sulphuric acid added, a precipitate would show that the barium had been present in the form of a sulphate.

Tests.

Oxalic Acid.

This is known in the commercial world as acid of sugar. It is a poison much used by suicides in this country, and has been the cause of death, in not a few instances, by being taken in mistake for Epsom salts or sulphate of zinc. It is rarely used

P

Symptoms. by homicides, on account of its ready recognition by reason of its bitter taste. When very concentrated, the symptoms will be mainly those of an irritant; when more dilute, and in considerable quantity, symptoms referable to the nervous system will be prominent. When taken in a concentrated form, the symptoms commence at once. They are: a sour acid taste, followed by a burning sensation in the mouth and throat, extending down to the stomach; vomiting of a grumous and blood-stained matter, and cramps in the extremities. The pulse becomes imperceptible, the patient is collapsed and bathed in a clammy sweat, the breathing becomes spasmodic and irregular, and the patient may die in less than half an hour. When the poison has been less concentrated, there have sometimes occurred tetanic spasms, with death from suffocation. Sometimes, when the poison was still more dilute, the patient has slept to death as if under the influence of some narcotic. After a smaller dose, there will be symptoms of inflammation of the mouth, œsophagus, and stomach, lasting for a varying length of time, and followed by more or less complete recovery. In such cases, there may also be dyspnœa, a peculiar loss of voice, cramps, pains in the head and back, restlessness, numbness of the arms and legs, and a peculiar eruption, described by Christison, a dull-red mottling in circular patches. The binoxalate of potash produces just the same symptoms as oxalic acid. The smallest quantity that has proved fatal is one drachm, taken in the solid state by a boy, aged 16, who died in eight hours. Three drachms might be regarded as a fatal dose. Death has several times occurred in less than half an hour, and in one case is said to have taken place in three minutes.

Post-mortem appearances. Occasionally, the alimentary canal has been found natural. More commonly, the mucous membrane of the mouth, throat, and œsophagus is whitish and softened; the œsophagus has a boiled appearance, or its lining membrane is thrown into folds and abraded in places. The stomach contains a dark mucous fluid of gelatinous consistency and acid reaction. The mucous membrane is soft, pale, and brittle, and often abraded; sometimes it is congested, and stained by extravasated blood; perforation has been seen, but is decidedly rare. The intestines will usually be found more or less congested and contracted. The heart is sometimes almost empty, at other times it is full of dark fluid blood; the lungs generally, and the brain occasionally, being found congested.

There are three cautions to be borne in mind *Treatment.* in regard to treatment. One is that the use of the stomach-pump is not free from danger, and, if possible, should be avoided altogether; another, that warm water increases the solubility of the poison, and should not therefore be used as an emetic, or, if used at all, it should be given in such large quantities, and at such a temperature, as to ensure speedy vomiting; the third is that alkalies should not be given as antidotes, as the salts they form are as poisonous as the acid itself. The proper antidote is a saccharated solution of lime. Failing this, some magnesia or chalk, mixed into a sort of cream with milk, may be given; if nothing else be at hand, some plaster may be scraped from the ceiling, and administered with water. For the rest, the treatment would be symptomatic.

On superficial examination, the crystals of oxalic acid resemble those of sulphate of magnesia. The taste, however, of oxalic acid is so bitter that *Tests.* there should be no confusion. A solution is acid, and with nitrate of silver gives a white precipitate, soluble in nitric acid, which, when dried and heated on platinum wire, detonates, and completely disappears as a finely divided white vapour. Sulphate of lime (and all the salts of lime, but none so characteristically) gives a white precipitate, soluble in nitric and hydrochloric acids, but insoluble in any vegetable acid. The urine should always be examined for oxalate of lime, the possibility of a very slight quantity being due to disease being always borne in mind. Under the microscope the crystals are octahedral, dumb-bell shaped, round or oval. To some urine chloride of lime is added, in excess, and then ammonia till the solution is alkaline; the precipitate should then be filtered, washed, and treated with acetic acid, which will dissolve the phosphates. The precipitate will then consist of oxalate of lime, mucus, &c. This should be treated with dilute hydrochloric acid, and filtered; to the filtrate a dilute solution of ammonia is added, in excess, when the oxalate of lime will be gradually precipitated in a crystalline form, and can be collected, dried, weighed, and tested in the ordinary way. A very minute quantity of oxalic acid in the stomach might possibly be the result of eating rhubarb or sorrel, but the quantity under such circumstances would, at most, be infinitesimal. Oxalic acid will be found in the stomach, in cases of poisoning, partly combined with lime, soda, or ammonia, and partly free, unless all has been neutralised by the antidotes that have been used.

Tartaric and Acetic Acids.

Poisoning by either of these substances is so rare that it is hardly necessary to do more than mention their names, and state that they have been known to act as irritant poisons. An instance has been recorded where a woman tried to commit suicide by drinking vinegar. She was seen three hours after taking it, her countenance was wild, and pupils dilated, her body was covered with cold perspiration, and the breathing was hurried and laboured; her tongue was dry and cold, the abdomen was distended, she had acute pain in the epigastrium which was increased by pressure, and was delirious. After an emetic of sulphate of zinc she made a speedy recovery.

CHAPTER XXI.

PHOSPHORUS, ARSENIC, ANTIMONY, ETC.

Phosphorus.

PHOSPHORUS is not often used by homicides, but has frequently been taken with suicidal intent. Many accidental deaths of young children have occurred from sucking lucifer matches, but as its use has, to a considerable extent, been given up in the manufacture of matches, there is reason to hope that the number of accidental deaths will be proportionately diminished. Phosphorus is a common ingredient in the pastes sold as vermin killers, and in the pure state is sold either in semi-transparent sticks, or in an amorphous condition, when it is red, and devoid of poisonous properties.

Symptoms. As a rule, the symptoms set in speedily, and consist of the taste and smell of garlic (the odour of which can be recognised in the person's breath); violent and intense pain in the mouth, throat, and gullet, pain at the epigastrium, nausea and vomiting. The vomited matters are luminous in the dark, and may be blood-stained; purging is often present. In very rapid cases, and especially in infants, delirium and coma may carry off the patient in a few hours; more commonly, within three days, the patient becomes jaundiced, and has retention of urine, which is bile-stained and albuminous. There is sleeplessness, with headache and frequent vomiting. Other nervous symptoms are numbness and tingling in the limbs, cramps, faintings, somnolence, delirium, trismus, and convulsions, followed by death. Or a series of hæmorrhages may take place into the skin, mucous membranes and viscera; this state may become chronic, and the patient may die from exhaustion after some months. Complete recovery may take place. Death usually occurs within a week, in one case it took place in four hours. In one instance a woman died after taking less than a grain in divided doses.

In a few instances no discoverable lesions have been present. The skin may be jaundiced, and the body surface, mucous membranes, and viscera may present scattered ecchymoses. The contents of the stomach and intestines, if they still contain the poison, may be luminous in the dark. The mucous membrane of the stomach will be found to be swollen, and its epithelium undergoing fatty degeneration; the swelling is due to enlargement of the tubular glands, which are blocked with degenerated epithelium; the same changes may be seen in the intestines, and there may be extravasations into the peritonæum. The liver shows the most characteristic lesions; it is large, yellow, has a mottled appearance on section, and is somewhat bloodless. Microscopically the hepatic cells have lost their well defined outline, and become filled with oil globules, the nuclei however being but little affected. The kidneys are generally swollen and engorged, and the epithelium of the convoluted tubes is found to be undergoing fatty degeneration. The muscular fibres of the heart also show a considerable amount of fatty change, and similar changes may be present throughout the whole muscular system. *Post-mortem appearances.*

The only disease, in respect of which any doubt could arise, is acute yellow atrophy of the liver. The latter is characterised by the early appearance of jaundice, by the decrease of the area of liver dulness, by the early appearance of nervous *Diagnosis.*

symptoms, and the rapid course of the disease. In phosphorus poisoning leucin and tyrosin may be present in the urine, but not to the same extent as in acute yellow atrophy. In the latter the liver is small, and its structure indistinct; in the former the liver is large, and the acini are distinct, the cells may contain fat globules, but are not granular.

Treatment. Oil of turpentine in full doses is considered to be an antidote, but some objection has been taken to its use, on the ground that phosphorus is partly soluble in it. It would be proper to give emetics, and afterwards solutions containing magnesia, and mucilaginous drinks; in other respects the treatment would be conducted on general principles.

Tests. The contents of the stomach, or any organic mixture suspected to contain phosphorus, should be warmed in the dark, when phosphorus will be shown by its luminosity and odour. There are three well known tests, viz.: 1. Mitscherlich's process.—This is based on the observation of the luminous properties of phosphorus vapour during condensation. The substance to be examined, which must be quite free from alcohol or ammonia, must be diluted till quite fluid, and then boiled, and distilled in a dark room through a cool tube. If a stream of water be kept running through the condenser, a luminous ring appears in the upper part of the tube. This is a very delicate test, but if the person has survived some days it will not be available, as it will be unlikely that any free phosphorus will be found. 2. Scherer's test.—The suspected substance is put in a flask, a little acetate of lead and some ether being added, a strip of filter paper soaked in nitrate of silver is then suspended in the flask, and the whole placed in the dark; after a few minutes the paper will be stained black, from phosphide of silver, if any phosphorus was present. 3. Blondlot's method.—This is dependent on the formation of phosphuretted hydrogen. Some pure hydrogen, obtained by the action of sulphuric acid on zinc, is passed into a flask containing the suspected substance; when this flask is heated a vapour will be given off, which may be recognised by means of the spectroscope, and also by its burning with a greenish flame.

The persons most liable to suffer from chronic poisoning are those engaged in the manufacture of lucifer matches, the makers of phosphorus paste, and chemists. Fortunately matches are now mostly made with allotropic phosphorus, which is not poisonous, and cases of poisoning are, therefore, becoming much less common. Dr. Bristowe investigated the subject for the Privy Council some years ago, and gave the following graphic description of the disease.

Chronic poisoning.

It begins usually, he says, with aching in one of the teeth; this is probably mistaken for an ordinary toothache. Sooner or later the tooth has to be extracted, and the pain for a time probably ceases. The wound in the gum, however, does not heal; offensive matter begins to ooze from it, and ere long a portion of the alveolus becomes exposed. The disease continues to spread, and, sometimes slowly, sometimes rapidly, more and more of the jawbone becomes denuded; the gums grow spongy and retreat from the alveoli, the teeth get loose and fall out. And thus the disease continues to progress, till in the course of six months, a year, two years—perhaps even five or six years—the patient sinks from debility, or from phthisis, or from some other consequence of the local affection; or, having lost piecemeal, or in the mass, large portions, one-half, or even the whole of the upper or lower jaw, he returns to his original state of good health, but the victim of a shocking and permanent deformity.

Iodine.

Acute poisoning by iodine is rare, and usually results from accident. In one case, for instance, tincture of iodine was swallowed in mistake for infusion of senna. The symptoms are those of an irritant, viz. burning pain in the mouth, extending down the gullet into the stomach, with the peculiar disagreeable taste of iodine in the mouth, nausea, pain in the abdomen, vomiting, and purging; the vomit and fæces being yellowish, or else of a dark blue or black, from the presence of starch in the digestive tract. The urine is suppressed, and there may be convulsions. The symptoms resulting from a too prolonged use of iodine are wasting, headache, fever, salivation, heat and dryness in the throat, vomiting, diarrhœa, diminution in the size of the mammæ or testicles, and enlargement and tenderness of the liver. Half a grain, taken three times a day, has been known to produce unpleasant symptoms. Inflammation of the stomach and intestines are found post mortem. Enlargement of the liver is said to be tolerably constant. Emetics may be given. The stomach should be thoroughly washed out. Starch or any farinaceous substance may be given. A colourless solution may be tested for iodine with some starch solution, when a blue colour will indicate the formation of iodide of starch. This test must always be performed cold. Organic mixtures, such as the contents of the stomach, should be shaken up with ether or chloroform, and, after being allowed to settle, the supernatant fluid decanted off, and tested with starch, or they may be shaken up with bisulphide of carbon, which gives with iodine a pink colour.

Iodide of Potassium.—Serious and alarming

symptoms have sometimes resulted from the use of this drug, but the only fatal cases have been when it was used in combination with chlorate of potash, iodate of potash being thus produced, which is an irritant poison. There is great difference between individuals in regard to their tolerance of this drug. The symptoms, produced by its too prolonged use, are coryza, salivation, ocular catarrh, gastric disturbances, and a peculiar vesiculo-pustular eruption, especially on the face and other parts which are commonly subject to acne, the irritant effects of the drug being manifested in the sebaceous follicles. Iodide of potassium is turned brown by sulphuric or nitric acid, and when heated gives off violet fumes. With starch solution and a little chlorine, a blue colour is given. Perchloride of mercury in solution gives a brilliant red precipitate, soluble in excess of either reagent.

Bromine.

The constitutional effects, produced by the excessive use of bromine, very closely resemble those due to iodine, sneezing, naso-ocular catarrh, and hoarseness being the most prominent; wasting of the mammæ or testicles is also liable to be caused. Only one fatal case has been recorded. Death occurred in seven hours after an ounce of bromine had been taken. The stomach contained four ounces of a fluid having a smell of bromine, and a reddish colour. Bromine may be tested for in the same way as iodine; the bromide of starch is yellow.

Bromide of Potassium.—The symptoms of chronic poisoning, by this drug, are very similar to those produced by iodide of potassium: running at the eyes and nose, and a peculiar and characteristic eruption, in some respects closely resembling acne. Microscopically, the changes in the affected skin are : 1. Hyperæmia of the corium, with exudation of corpuscles, especially in the neighbourhood of the papillæ. 2. The formation of collections of pus corpuscles, or minute abcesses, in the vicinity of the hair follicles and sebaceous glands. In the only fatal case on record from the administration of bromide of potassium, death appeared to be due to syncope.

Arsenic.

White arsenic.

Metallic arsenic, when pure, is not poisonous, but it is readily oxidised. Arsenious acid, or white arsenic, is the most important compound. It occurs as a white crystalline substance, or as a white amorphous powder, and is not very soluble in water, even when this is boiling. It has a slightly sweetish taste. The sale of arsenic to the public is under the restrictions of the Poisons Act, and it is mixed with $\frac{1}{6}$ of its weight of soot, or $\frac{1}{32}$ of its weight of indigo, before being sold. Thus it happens that in a case of poisoning by arsenious acid, the vomited matters may be white, blue, black, or yellow from the admixture of bile. The presence of organic matter in a liquid renders the poison less soluble; the presence of an alkali, or an alkaline carbonate, greatly increases its solubility. A very large quantity can be held in suspension in viscid fluids, *e.g.* gruel, arrowroot, cocoa, &c.

Symptoms.

The symptoms appear in about half an hour after the poison is taken, but they may come on almost immediately, and in very exceptional cases they may be very much delayed. In the most acute and rapidly fatal form, there is burning pain

Irritant form.

in the mouth, throat, gullet, and stomach, with sometimes an acrid taste in the mouth; nausea, and vomiting, appear early; there is great thirst, and often hoarseness. The abdomen becomes tender, and sometimes swollen; and the vomiting is very violent. Diarrhœa is usually present, with tenesmus; in some cases, when the diarrhœa is very severe, the disease may resemble cholera. With the diarrhœa there is also burning pain in the intestines, and excoriation of the anus. Among other symptoms there may be difficulty of breathing, and a sense of tightness about the lower part of the chest. Frequent, difficult, and painful micturition is sometimes noted. Death may occur from collapse within twenty-four hours. More commonly the course of the disease is not so acute, the same symptoms being present, but in a less aggravated form. In this form, remissions towards the end of the second day, or a little later, are not uncommon; and nervous symptoms are apt to appear, such as cramps in the limbs, especially the calves, and tremors and twitching all over the body; death may occur in a paroxysm of convulsions. In other cases the nervous phenomena predominate, loss of power in the limbs gradually becoming complete; delirium, convulsions, and stupor, deepening into coma, being the main features. In other cases again there are hardly any signs of irritation or inflammation, faintness being the chief or only symptom. In cases that do not prove fatal, the nervous symptoms generally last after all the inflammatory symptoms have passed off, local paralysis in the arms or legs being a not uncommon sequel, and persisting for some time. *Nervous form.*

The following instance of wholesale accidental poisoning, in a mild degree, is worth quoting. At a large industrial school near London, in December 1857, three hundred and forty children were suddenly *Case of wholesale accidental poisoning* seized with symptoms of poisoning by arsenic, soon after breakfast. They had been supplied with milk diluted with water from a boiler, into which, to cleanse it of fur, a quantity of an alkaline solution of arsenic had been placed. Upon an average, each child took a grain of arsenic. The symptoms varied but little. There was shivering, with pain in the stomach and bowels, and in most of the cases vomiting of a clear, ropy, mucous fluid, of a green colour. These symptoms were developed within one hour. About three hours after the meal, pain in the forehead, more or less intense, was a marked symptom, and there was a copious discharge of a watery, colourless fluid from the nose. Seven had cough of a croupy character, three vomited blood, and one passed blood by the bowels. Some suffered from severe inflammation of the stomach; of these only six were under treatment at the end of the first week, and one did not recover till after the second week. All the children recovered.

Death has been known to be caused by two grains. Great tolerance of the drug can be established by habit, as in the well-known instance of the Styrian peasants, who have been known to consume as much as six grains at once without any ill effects. In one instance death occurred in twenty minutes.

Poisoning by small and repeated doses. Sometimes, in homicidal cases, the poison has not been administered all at once, but in gradual doses, so as to avoid any suspicion, and encourage the belief that the illness was due to some natural cause. Vomiting, especially after meals, and emaciation would be the most prominent symptoms in such an illness. Amongst other symptoms that may appear when the poison is administered in this fashion, are conjunctivitis, intolerance of light, and a vesicular eruption known as eczema arsenicale. Numbness and tingling in the limbs, and paralysis may also be complained of. Falling off of the hair and salivation have been noted. The secret poisoners of the middle ages used arsenic to get rid of their victims; the most notorious was a woman of the name of Toffana, who, by means of her Acquetta di Napoli, is credited with having poisoned six hundred persons.

In a few cases, especially those where death has supervened early with cerebral symptoms, or from heart failure, no morbid appearances have been found, but such are quite the exception. The stomach is the organ chiefly affected, this being the case even when the poison was not administered by the mouth. The mucous membrane of the stomach is deeply congested, showing here and there ecchymoses; ulceration or perforation is decidedly rare, the redness is often more marked at the greater curvature. The membrane may appear smeared as with a thick white or yellow substance, this last colour being due to the formation of a sulphide of arsenic. The duodenum is usually affected, and often the whole alimentary canal is inflamed; the rectum rarely escapes. The liver and kidneys show a certain amount of fatty change, when examined under the microscope. The spleen may be swollen and soft, and the heart and bladder may show ecchymoses beneath their lining membranes. The mouth, tongue, and gullet may show traces of inflammation. The blood is generally fluid. Arsenic possesses, in a high degree, the property of preventing putrefaction; this was well shown at a recent trial in Liverpool. The body of a man who had died on December 7th, 1880, was examined on January 16th, 1884, when the viscera were found to be very much shrunk, but well preserved, and emitted very little odour. Some caution however is necessary, as Casper has pointed out, in attributing redness of the mucous membrane, found some time after death, to arsenic,

Post-mortem appearances.

as there is a redness which occurs as a result of putrefaction. Arsenic is used by bird stuffers, owing to its antiputrefactive properties.

In the case of a medical man, who died in sixteen days from taking arsenic accidentally, none of the poison could be detected in the body; it is not usual, however, for complete elimination to have taken place so early as this. Arsenic is found in the liver for a longer time than in any other organ. There are many ways in which death has been caused by the external application of arsenic, either through the unbroken skin as in the use of violet powder, or through an abraded surface as in the treatment of ulcers of the leg, cancerous tumours of the breast, &c. The introduction of arsenic, too, into the vagina or rectum has been followed by death.

Treatment. Treatment should never be neglected because the case seems hopeless. If the patient be seen quite early, the stomach-pump might be used; as a rule vomiting sets in spontaneously, but if there is any necessity emetics may be given, and after vomiting, milk and albuminous drinks. Ferric hydrate is an antidote to arsenious acid, but it will be of no use to give it after the poison has been absorbed. It acts by converting arsenious acid into an insoluble ferric arseniate, but it is essential that the ferric hydrate be quite fresh. It is easily prepared by adding strong ammonia to the tincture of the perchloride of iron. Opiates, chloroform, or venesection might be indicated in any particular case. If the patient can swallow he should be made to drink large quantities of water, in order to eliminate the poison by the kidneys.

Tests. The tests for arsenic will be found in all the chemical text-books, and it is unnecessary to recapitulate them all here. The two that are best known and most valuable are those of Reinsch and

Marsh. The former is very simple. The solution supposed to contain arsenic, should be rendered acid by the addition of hydrochloric acid, and then a piece of clean copper foil, or copper wire, is placed in it, and the whole boiled for some minutes. If any arsenic be present it will form an iron-grey coating on the copper. This may be washed and dried, and then tested for arsenic. Marsh's test depends upon the formation of arseniuretted hydrogen. If a solution containing arsenic be placed in a wide-mouthed beaker, and some zinc and sulphuric acid added, arseniuretted hydrogen gas will be disengaged. The purity of the zinc must be ascertained before the test is commenced, as arsenic is one of the impurities of zinc; this is easily done by testing the gas obtained from the action of the sulphuric acid on the zinc alone, this gas should be pure hydrogen. Arseniuretted hydrogen is a colourless vapour having an odour of garlic. It may further be recognised by the following test. The gas should be conducted from the flask, where it is generated, through a glass tube bent at a right angle outside the flask, and drawn out to a fine point; as it escapes the gas can be ignited, and if a piece of cold porcelain be held in the flame a deposit of metallic arsenic will form upon it. A solution containing antimony, when similarly treated, will also give a gas which will burn, and cause a stain, the characters of which must be closely studied in order to distinguish it from the above. The arsenical stain is brilliant and of a hair-brown colour; it is volatile and easily dissipated by heat. If the plate be warmed, and a current of sulphuretted hydrogen be passed over the stain, this will become yellow, owing to the formation of the sulphide. Chloride of lime completely dissolves the stain. Sulphide of ammonium does not dissolve it, but on heating, the stain becomes yellow with a metallic

226 MEDICAL JURISPRUDENCE.

centre. The characters of the antimonial stain will be referred to subsequently.

The limit for the complete elimination of arsenic from the body is one month. Of all the viscera the liver shows traces of arsenic the longest. Arsenic may be detected in the dead body, even ten years after death. When the coffin has given way it is necessary to examine the soil surrounding as well, as this may contain arsenic; but in this case the arsenic exists as the oxide, a highly insoluble form, so that, if a solution obtained by water from any part of the body yielded evidence of arsenic, it would be quite safe to assert that the arsenic came from the body and not from the soil. The grave clothes should also be examined, as, if the poison was due to the soil, they would be impregnated before the body.

Compounds of arsenic. There are many preparations containing arsenic which have, at one time or another, proved dangerous and therefore require mention. Amongst them is Arsenite of Potash, or Fowler's Solution. It acts in the same way as arsenious acid. Arsenite of potash and soda is used by shepherds to kill the fly in sheep, and hence its name of fly-water. It is a poisonous compound. Arsenite of Copper, or Scheele's Green, contains one part of arsenious acid to two of oxide of copper, and is very poisonous.

Emerald Green. Aceto-Arsenite of Copper, or Emerald Green, demands more attention. It consists of six parts of arsenious acid, two of oxide of copper, and one of acetic acid. It has been largely used, in the manufacture of wall-papers, because it is a 'fast' colour.

Arsenical wall-papers. A green paper is always to be regarded with some suspicion, but all green papers do not contain arsenic, nor is the absence of a green colour any guarantee that the paper is free from arsenic, since this has been detected in dark brown, buff, white, blue, drab, mauve, and various shades of grey papers. An unglazed arsenical paper is highly dangerous; varnishing and glazing diminishes the danger. The best manufacturers of the present day, however, use no arsenic at all. In France, Germany, and Sweden, there are most stringent regulations against the use of poisons, and of arsenic in particular, in the manufacture of various domestic articles. The people most liable to suffer are the papermakers, the paperhangers, and those who inhabit rooms where such papers are in use. If the wall-paper contains arsenic, the dust is sure to contain the poison, the air in such a room will also contain minute traces of arseniuretted hydrogen. Arsenic is also used in the manufacture of mordants for calico printing and dyeing, in the manufacture of cretonnes, and in the preparation of aniline dyes. Arsenic may also be found in the bright colours on children's toys, in various articles of confectionery, syrups, jellies, wines, and the glazed wrapping papers used for lozenges, &c. As in the case of wall-papers, so in regard to these substances, the absence of a green colour does not prove that the article, in question, is free from arsenic. It is to be remembered that it is not because they are green that the things are poisonous, for the green colour is due to the copper and not to the arsenic.

The symptoms of chronic arsenical poisoning, thus induced, do not differ from those already mentioned, irritation about the eyes, dryness of the throat, coryza, and general languor being the earliest. Headache, nausea, cramps, pain and tenderness in the abdomen ensue, the hair falls out, there may be painful micturition, the person becomes cachectic, and sometimes jaundiced. In paperhangers there is a peculiar eruption about the root of the nostrils, backs of the ears, bends of the elbows, insides of the thighs and scrotum, *Symptoms of chronic poisoning.*

papular at first, afterwards becoming pustular, leading to the formation of circular superficial ulcers; the nails may come out. To examine a suspected wall-paper, dip some of it into a strong solution of ammonia, if it becomes blue then a copper salt is present, pour some of this blue liquid over a crystal of nitrate of silver in a white dish, a yellow colour would show the formation of arsenite of silver.

Arsenic acid is said to be a very deadly poison, but no case of death from its use appears to have been recorded, it is a white, uncrystalline, deliquescent solid. The arsenates of potash and soda are used in the manufacture of aniline dyes, and are highly poisonous. The sulphides of arsenic are two, viz. realgar or red arsenic, and orpiment or yellow; when quite pure they are not poisonous, but they are readily oxidised, and then become very poisonous from admixture with arsenious acid. Arseniuretted hydrogen or arsine, to which allusion has already frequently been made, is a very poisonous gas. Gehlen, the discoverer of it, died in eight days from the effects of inhaling the fumes. Nine persons, employed in a factory, were exposed to the fumes for a varying length of time, three died, and of the rest all but one (who was very little exposed) were unable to resume their work for some months. The symptoms consist of malaise, nausea, vomiting, a red and furred tongue, pains in the limbs, hæmaturia and jaundice. The blood assumes an inky colour.

Antimony.

Tartar emetic.

There are only two salts of antimony of much interest from a medico-legal point of view, viz. potassio-antimonio-tartrate and chloride of antimony. The former is a whitish crystalline powder, not readily soluble in water. It is not commonly used as a poison, but has occasionally been taken in mistake for cream of tartar.

A highly metallic taste is at once perceived on swallowing tartar emetic, and lasts for some time. Nausea and vomiting appear early, then pain in the abdomen and purging. Heat and constriction of the throat are complained of, and a whitish crust appears in the mouth. Cramps supervene in the limbs, cold sweats, great faintness and weakness; the pulse becomes very feeble, the eyes sunken, and the voice reduced to a whisper. Death may take place from convulsions, or the mind may remain clear to the last. The symptoms are very like those from arsenical poisoning, but there are no remissions; the urine is not so likely to be suppressed, and the general depression is more marked. Two grains have proved fatal to an adult, and smaller doses have been fatal to persons in bad health. Sometimes the poison is administered in small and repeated doses; in such a case nausea, vomiting after the taking of the poison, and pain and distension of the abdomen would be prominent symptoms. There would be gradually increasing weakness with purging, and perhaps blood in the motions. As in the case of arsenic, a pustular eruption is liable to appear. Occasionally poisonous symptoms have been seen from its external application, but such cases are rare; it readily, however, produces a pustular eruption. *[Symptoms.]*

In the well-known case (which may be taken as a typical instance) of Mr. Bravo, who died from a dose that probably exceeded 40 grains, and who survived for 55½ hours, the stomach contained thick gruel-like matter, no inflammation or ulceration was found; the duodenum was soft, its surface was pale and yellowish, there was no ulceration. At the lower end of the ileum were some red spots, with a yellow pasty matter, the cæcum *[Post-mortem appearances.]*

contained several small ulcers; the ascending colon was deep red and contained clots of blood, the rest of the colon was deeply blood-stained, and contained fæcal matter mixed with blood. The other viscera showed no important changes. As a general rule the heart is empty, and the blood fluid throughout the body. The liver may be enlarged and show fatty changes.

Treatment. Vomiting must be encouraged in every way, the use of the stomach-pump being permissible; plenty of warm milk and water should be given. Any astringent infusions may be given as antidotes, such as tannic acid in solution, oak galls, or even strong green tea. For the rest the treatment would be symptomatic.

Tests. With Marsh's test the stain on a piece of cold porcelain is distinguished from that of arsenic by the following characters: It is sooty, and not so volatile. It gives no precipitate with nitrate of silver after being dissolved in nitric acid. It is not dissolved by hypochlorite of lime. It is dissolved by sulphide of ammonium. When the tube, through which the gas is passing, is heated, the deposit of the metal, as a black powder, takes place on either side of and quite close to the flame, whereas in the case of arsenic the metal is deposited only beyond the flame, and at some little distance from it. With Reinsch's test the deposit on the copper has a more sooty appearance than in the case of arsenic. If the copper with the deposit on it be boiled with a solution of caustic potash, the resulting liquid can be tested for antimony by any of the ordinary tests, which will be found in the chemical text-books.

Chloride of antimony. The chloride of antimony, or butter of antimony, is a very corrosive fuming liquid. A case has been recorded in which it proved fatal in two hours. The symptoms are those of an active and violent

ANTIMONY.

poison. In the case of a man who died $10\frac{1}{2}$ hours after swallowing two ounces, the interior of the alimentary canal, from the mouth to the jejunum, presented a black appearance as if the parts had been charred; the mucous membrane was destroyed, being represented only by a flocculent substance, which could easily be scraped off from the tissues beneath. If chloride of antimony be added to water a yellowish-white precipitate of oxychloride is thrown down; this precipitate can be shown to contain antimony by any of the ordinary tests, the filtrate can be shown to contain hydrochloric acid by the addition of nitrate of silver.

[handwritten notes:]
+ H_2S = orange red
+ Ferro Cy of Pot = no precip. } *Littlejohn*
+ H_2SO_4 dilute = white soluble in excess

CHAPTER XXII.

COPPER, MERCURY, LEAD, ETC.

Copper.

ALL the salts of copper are poisonous, but the most important are the sulphate or blue vitriol, the carbonate or natural verdigris, and the subacetate or artificial verdigris.

Symptoms. A coppery taste in the mouth, violent vomiting and retching, constriction in the throat, thirst, and pain and tenderness in the abdomen are the most prominent symptoms. The vomited matters may be blue or greenish, purging is generally present, with tenesmus, the fæces being the same colour as the vomit, or sometimes discoloured by blood. Jaundice and suppression of urine are both tolerably constant. In the later stages certain nervous symptoms make their appearance, such as cramps, spasms, paralysis; the breathing is hurried and difficult, the extremities cold, and the patient may die comatose or with tetanic symptoms. If the vomit is greenish a very simple test will determine whether this is due to bile or not. Add a few drops of ammonia, a deep blue colour would indicate the presence of copper.

Post-mortem appearances. The mucous membrane of the stomach and intestines will be inflamed and thickened, and sometimes has been found ulcerated; it may be discoloured to a dirty bluish green from contact with the salt. The liver is often friable and fatty, and the gall bladder nearly empty. The kidneys may be swollen, especially the cortex which may also be pale and bile-stained.

Emetics should be given, and the stomach-pump freely applied. Substances containing albumen form an insoluble compound with the oxide of copper, on which the acids of the stomach have no action, and therefore white of egg and milk should be given. Sugar has been recommended on the ground that it reduces salts of copper to a sub-oxide, which is nonpoisonous, but it is open to doubt whether this change would take place in the stomach. *Treatment.*

Salts of copper when heated in the inner blow-pipe flame give a green colour to the outer flame. Sulphuretted hydrogen, passed through an acid solution containing copper, gives a greenish-black precipitate soluble in boiling hydrochloric acid or in boiling nitric acid, when a blue colour results. Ammonia gives with concentrated solutions a precipitate soluble in excess, when the liquid becomes of a deep blue colour. Ferrocyanide of potassium gives a reddish-brown gelatinous precipitate, or with dilute solutions a rich claret colour, but no precipitate. A clean needle inserted into a solution of copper becomes coated with a thin reddish metallic film, the iron being slowly dissolved. *Tests.*

The symptoms of chronic poisoning are those of general muscular debility, a coppery taste in the mouth, obstinate colic, constipation or diarrhœa, and jaundice. Some have described a green line on the gums. The late Sir Dominic Corrigan, who paid a good deal of attention to the subject, described a characteristic retraction of the gums with a purple (not blue) edge. *Chronic poisoning.*

Amongst the causes of this form of poisoning are the following:—Employment in trades where copper has to be filed and handled, or where sul- *Its causes.*

phate of copper exists in a finely divided state, and so is likely to be inhaled. Its use in baking to give the bread additional whiteness. Its presence in the water in the neighbourhood of copper mines. Its use to colour confectionery, pickles or preserved fruits; of late years there have been several prosecutions for the adulteration of tinned peas by copper. The use of copper vessels for culinary purposes. This last is the most important of all, and has been a fertile source of this form of poisoning. It has been suggested that the attacks of cholera, and of acute or chronic dysentery, from which Europeans so frequently suffer in tropical climates, are, in many cases, due to the general employment of copper utensils for culinary purposes, and to want of cleanliness on the part of native cooks, who use butter, salt, and acids without removing the cupreous incrustation which is formed on the surface or on the rims of the vessels. Hot butter, lard, or hot oil readily dissolve copper, forming fatty acids of which oxide of copper is the base.

Within the last few years the use of copper vessels for ordinary culinary purposes has been very generally discarded. The reasons for this are obvious. Distilled water when boiled in a perfectly clean copper vessel, if air be excluded, has no action on the copper, but if air be present some of the copper will be dissolved. Acid solutions, such as vinegar, or acid wines, and fatty and oily matters (especially the volatile oils and rancid oils), when boiled in copper or brass vessels and allowed to cool, will dissolve some of the copper. If the vessel used be perfectly clean, and if the substance be poured out when hot, and not allowed to stand to cool in the vessel, no copper will be dissolved. Milk, tea, coffee, beer, and water containing cabbages, potatoes, turnips, carrots, rice, onions, and barley have no action on copper.

Mercury.

Metallic mercury is not usually regarded as a poison when in the solid state, and, half a century ago, it was frequently administered in enormous doses in cases of intestinal obstruction (occasionally with very good results) merely for a mechanical effect. Thus a boy, aged ten, took three ounces of mercury on two occasions with a day's interval, and completely recovered, all the mercury being passed by stool except half a drachm. A remarkable instance of wholesale poisoning by mercurial vapour occurred on board a man-of-war, the 'Triumph,' in 1810; when some leathern bags containing mercury burst, and three tons of mercury were dispersed throughout the vessel. The crew soon began to suffer from salivation, partial paralysis, and diarrhœa, and in three weeks 200 men were salivated.

Corrosive sublimate, or perchloride of mercury, is the most important of the salts of mercury. It is a heavy crystalline substance, or may occur as a powder; it is very soluble in hot or cold water, and in alcohol. <small>Corrosive sublimate.</small>

At the time of swallowing the poison there is a metallic taste in the mouth with constriction in the throat, then burning pain extending down the gullet to the stomach, the mucous membrane lining the mouth becomes whitish and shrivelled. There is nausea, vomiting of mucus mixed with blood, and shortly purging of a similar character, and there is pain at the epigastrium increased by pressure. The patient becomes more or less collapsed, the pulse being scarcely perceptible, the urine is scanty or suppressed. Salivation is not common as a result of acute poisoning. The symptoms differ from <small>Symptoms.</small>

those in arsenical poisoning chiefly in the metallic taste, in coming on sooner, and in the more frequent presence of blood in the vomit and fæces. Three grains have killed a child. Death in one case took place in half an hour; the average fatal period would be from three to six days. Corrosive sublimate is readily absorbed through the skin, and produces the same constitutional disturbance as when taken into the stomach. This fact is daily made use of in practice. The limit for complete elimination of mercury, after inunction, is probably greater than three months.

Local application.

The whitened condition of mucous membrane, to which allusion has already been made as regards the mouth, will be found also in the œsophagus. The mucous membrane of the alimentary canal often shows signs of inflammation throughout, ecchymoses being common in the stomach. The cæcum and colon may be highly inflamed, and are sometimes ulcerated. The kidneys usually show some inflammation, and the bladder may be empty, contracted, and its lining membrane somewhat inflamed. As in the case of arsenic, when applied locally, mercury produces inflammation of the stomach.

Post-mortem appearances.

Vomiting should be encouraged by the administration of large quantities of water and mucilaginous drinks, distilled water is particularly recommended; the use of the stomach-pump would not be without risk, owing to the state of the mucous membrane. White of egg, milk, and gluten would constitute the best antidotes.

Treatment.

The salts of mercury are readily volatilised, and, when heated in a tube with some carbonate of soda, a sublimate of metallic mercury is deposited in a cool part of the tube. If a solution containing iodide of potassium be slowly added to a solution of corrosive sublimate, a yellowish pre-

Tests.

cipitate will be thrown down, gradually changing, on the addition of more iodide, to a vermilion; this precipitate is soluble in excess of either reagent; this constitutes a simple, delicate, and reliable test. Proto-salts of mercury give a black precipitate with sulphuretted hydrogen, or sulphide of ammonium; persalts, under the same circumstances, give at first a whitish one, gradually becoming black; this precipitate is insoluble in boiling nitric acid. If a piece of bright copper foil be dipped into a solution of corrosive sublimate, previously acidulated with hydrochloric acid, a whitish deposit of mercury will form upon it, which may be scraped off and tested.

There are two well-marked secondary, or chronic, effects of mercury which call for notice, viz., salivation and tremor. Of these the former is more likely to be set up by repeated small doses than by one large one. It is accompanied by a peculiar fœtor of the breath, swelling of the gums and whole face; the gums often ulcerate, and the teeth fall out. In the worst cases there is a constant flow of saliva from the mouth, the face is swollen so that the person cannot be recognised; gangrene of the tongue or cheek may set in, and sometimes a large portion of the cheek sloughs away. It is to be distinguished from cancrum oris, in the early stages, by being general over the whole mouth, and by the affection of the tongue, whereas the disease commences on one side of the mouth, in the cheek; in late stages it might be possible that testing the saliva for mercury would be the only means of distinguishing the cause. Children are not so readily affected by mercury as adults. Persons suffering from Bright's disease are very susceptible to the influence of mercury. In the other form of chronic poisoning there is gradual weakness, followed by tremor, at first confined to the

Salivation.

Mercurial tremor.

arms, but afterwards involving the legs; the patient is very anæmic, and may become irritable or melancholic; and in later stages may be completely paralysed, his mental faculties also becoming impaired, and in this state he may die. The people liable to this sort of poisoning are quicksilver miners, water-gilders, mirror silverers, and barometer makers. They often have a blue or dark red line at the edge of the gums, and their teeth become brittle and easily chipped. Cleanliness on the part of the workmen, and free ventilation of the rooms where they work, are the best precautions that can be taken; the floor of the workshop should be sprinkled, after hours, with a solution of ammonia. Recovery is only possible where the patient gives up his occupation before the affection is well established.

The other salts of mercury are all more or less poisonous. They are calomel, white precipitate, red precipitate (mercuric oxide), Turpeth mineral (mercuric oxysulphate), vermilion (mercuric sulphide), nitrates of mercury, mercuric cyanide, Pharaoh's serpent (mercuric sulphocyanide), and mercuric methide. Of these the persulphide has been used by dentists as a colouring matter for artificial plates, and, being constantly in the mouth, has been acted upon by the saliva, and so has given rise to symptoms of poisoning. Two laboratory assistants lost their lives some few years ago by being poisoned whilst engaged in the manufacture of mercuric methide, a compound of mercury, which may be regarded as corresponding to the perchloride, in which the chlorine is replaced by methide. In both the central nervous system was affected.

Lead.

Lead when pure is not poisonous, but it is easily oxidised. All the salts of lead are poisonous, the most important being the acetate, or sugar of lead, the chief cause of acute poisoning, and the carbonate or white lead, the commonest cause of chronic poisoning.

In acute poisoning, which is far from common, the chief symptoms are a metallic taste, with dryness and burning in the mouth and throat. Vomiting soon sets in, with pain in the abdomen, and, as a rule, constipation, the motions being dark coloured. The pulse is small and frequent, the skin dry, the urine scanty, cramps and numbness in the limbs appear with, perhaps, convulsions; death is decidedly rare, though, in a recent case, an old woman was murdered by the repeated administration of small doses of acetate of lead. A wholesale case of poisoning occurred some years ago, when a large quantity of the acetate of lead was mixed with flour, and made into bread; over five hundred persons presented symptoms of poisoning, but all recovered. The stomach and intestines are lined with a thick layer of mucus, whitish or whitish yellow, from admixture with the salt of lead; beneath this the mucous membrane shows traces of inflammation. But there might be nothing very characteristic. In chronic cases, besides the blue line on the gums, a deposit of lead will be found in the submucous tissues of the intestinal walls. *[Symptoms. Post-mortem appearances.]*

The best antidote would be the sulphate of magnesia, as a very soluble sulphate, ready to combine with any lead in the alimentary canal, and form the highly insoluble sulphate of lead. *[Treatment.]*

Tests.

Solutions containing lead give a black precipitate with sulphuretted hydrogen insoluble in caustic alkalies or dilute mineral acids. Sulphuric acid gives a white precipitate soluble in boiling hydrochloric acid. If a piece of zinc be suspended in a solution of lead, metallic lead will be deposited upon the zinc. Water may be examined for lead by simply passing sulphuretted hydrogen gas through it; if there is no blackening, there is no lead.

Chronic poisoning.

Persons who suffer from chronic poisoning by lead are apt to become emaciated and cachectic. They suffer from anæmia, are constipated, and become dyspeptic. The most constant sign is the presence of a blue line on the gums; this is rarely absent, and is pathognomonic. It is of a dark colour, and occurs along the margin of the gums next to the teeth. It only occurs opposite the teeth, and is most marked in cases where cleanliness is not observed. Microscopically it has been found that it is due to the deposition of a sulphide of lead in the capillaries of the gum. How soon it may be formed, or how long it may last, has not been ascertained as yet with certainty.

There are two distinct types of chronic lead poisoning, not necessarily occurring in one and the same person, though they may do so. Of these one is really acute in its symptoms and course,

Lead colic.

viz. *lead colic*, also called painter's colic, colica Pictonum (from its being common among the inhabitants of Poictou), dry belly-ache, and Devonshire colic (from its frequent occurrence in the cider manufacturing districts). It commences with obstinate constipation, and severe pain in the abdomen, which becomes so intense that the patient lies with his knees drawn up; the abdomen is hard and retracted, and the pain is usually relieved by pressure. Vomiting is commonly present; the skin is dry, but there is not often much fever; the urine is scanty. Death in an attack is quite exceptional. When the pain is not relieved by pressure, it is supposed that the deposit of lead has taken place in the muscles of the abdominal walls, whereas in the majority of cases it is deposited in the coats of the intestine. During the attack a hot bath (104° F.) seldom fails to give relief. A person who has once had an attack of colic, is liable to another if he does not avoid the causes which induced the first. The other form of chronic lead poisoning is that known as *lead palsy*. As a rule, this is a later result than the colic, but it by no means follows that a patient with lead palsy has previously suffered from lead colic. The muscular weakness shows itself most in the extensors of the forearm, which are wasted and very weak, producing the well-known and characteristic wrist-drop. The only disease, which could be mistaken for this, would be progressive muscular atrophy in a somewhat advanced stage, but, apart from the blue line on the gums and the electrical reactions, lead palsy is characterised by being more or less general from the first, whilst progressive muscular atrophy, almost invariably, begins in the ball of one thumb some time before it appears elsewhere. There is a rapid loss of faradaic excitability in the affected muscles, the response to the constant current remaining normal, or even being a little exaggerated. As yet no constant lesions have been discovered; fibroid degeneration of the nerves has been described, and a deposit of lead in the muscular fibres has been found, but the symmetry of the disease would strongly suggest a central origin.

Lead palsy.

Amongst the other associations, or sequelæ, of lead poisoning may be mentioned gout, granular kidneys, melancholia or mania, and epilepsy. It

R

242   MEDICAL JURISPRUDENCE.

is quite possible that these are due in great measure, as a recent writer has suggested, either to arterial spasm, or, in their more chronic manifestations, to arterial thickening. The causes of lead poisoning are very numerous, the most important being employment in a white-lead factory. House painters, gilders, type-founders, type-setters, and those whose work compels them to be habitually handling lead, are liable to it. The impregnation of drinking water by lead is certainly one of the most common causes, and a short time ago was the subject of an action in our law courts. Rain water, when stored in a leaden cistern freely exposed to the air, gives rise to the formation of an oxide of lead, which is dissolved in the water; this solution then absorbs carbonic acid from the air, and a precipitate of oxycarbonate is thrown down. In a hard water (one containing sulphates, phosphates, or carbonates of lime) this reaction does not take place, and the water is fit for drinking.

Action of water on lead.

Zinc.

The sulphate is well known for its emetic properties, and may be regarded as a mild irritant poison. Occasionally it has been taken in mistake for Epsom salts. The chloride is an active irritant and corrosive poison, and has caused death both when taken internally, and when applied externally, *e.g.* to a cancerous breast. Pain and burning in the mouth, throat, and epigastrium are the first symptoms; dysphagia, increase in the quantity of saliva, vomiting and purging, sometimes of blood, cramps, collapse, and coma followed by death. Sometimes, when the dose has not proved immediately fatal, the patient has died, some months later, from stricture of the œsophagus. Burnett's disinfecting fluid owes its properties to

Symptoms.

this salt, and has often caused death. In acute cases the mucous membrane of the mouth and œsophagus may be congested, but sometimes is white and opaque, being at the same time tough and resembling leather in consistence. Perforation has been recorded in one case, where a woman, aged sixty-three, died fifteen hours after taking an ounce and a half of Burnett's solution. The œsophagus was found in shreds at its lower third. Tepid water with milk or albumen should be freely given to promote vomiting; the stomach pump should be used with caution, bearing in mind the corrosive nature of the poison. Carbonate of soda might be given as an antidote. Salts of zinc, when heated on platinum foil, give a residue which is yellow when hot, white when cold. If a solution of zinc be placed in a platinum capsule, and the platinum be touched through the solution with a piece of magnesium, zinc will be deposited on the platinum. *[Post-mortem appearances. Treatment. Tests.]*

Iron.

Salts of iron have but rarely proved poisonous. Occasionally a poisonous dose has been administered with a view to abortion. A series of cases was reported from Martinique in 1876, where a woman had, on three separate occasions, poisoned men with the perchloride, death resulting in each case with symptoms of irritant poisoning. The contents of the alimentary canal, in such a case, would be of a greenish-black colour.

Silver.

The nitrate is the only salt that has any interest from a medico-legal point of view. Cases of poisoning by it are very rare, and generally occur in children as the result of accident. Thus an infant put a piece of stick nitrate of silver into its mouth and swallowed it. Vomiting and purging set in, followed by convulsions and death in six hours. Common salt would be the best antidote to give in such an emergency. Nitrate of silver is readily absorbed, and is gradually deposited in the viscera and tissues. When taken for a long time the skin becomes stained of a leaden hue, giving the person a very unnatural complexion; a purple or violet line along the gums has been described in these cases. Heated on charcoal with carbonate of soda, silver salts give a white malleable bead of silver and no incrustation. Chloride of silver, when thus heated, fuses into a soft greyish horny mass. In solution hydrochloric acid gives a dense, curdy, white precipitate. Caustic alkalies give a brown precipitate, soluble in excess of ammonia.

Bichromate of Potash.

This is an active irritant poison, producing vomiting and purging, thirst, collapse, and cramps in the limbs. Amongst other nervous phenomena there may be suppression of urine, paralysis of the legs, and dilatation of the pupils. The stools and vomit are generally of a yellow colour. Cases of acute poisoning are not common. Death has been known to take place in four hours after two drachms had been taken. This, however, should not be regarded as a necessarily fatal dose. Inflammation of the mucous membrane of the alimentary canal is the most constant lesion that has been found hitherto. For treatment, emetics of sulphate of zinc should be given, and afterwards carbonate of magnesia in milk. Those engaged in the manufacture of bichromate of potash are liable to suffer from its effects. These consist of a disagreeable taste in the mouth, and irritation about the mucous membrane of the nose, ulceration and gradual destruction of the septum nasi taking place; conjunctivitis and increased flow of tears are also amongst the symptoms. If the workman has any sores about his hands, feet, or shoulders, they are liable to take on a bad type of ulceration, forming large ulcers with hard edges; but it appears that these never form where the skin is sound. When heated in the borax bead a green colour results. Heated on charcoal with nitre and carbonate of soda a yellow mass is formed. In solution sulphuretted hydrogen gives no precipitate; sulphide of ammonium gives a green precipitate.

Chronic poisoning.

Tests.

CHAPTER XXIII.

VEGETABLE AND ANIMAL IRRITANTS.

UNDER the head of vegetable irritants would come the following:—Euphorbia, castor oil, jatropha, croton oil, elaterium, colocynth, bryony, ranunculus, anemone, stavesacre, marsh marigold, mezereon, savin, jalap, and hellebore. The symptoms produced by these are very similar, and are, mainly, those resulting from inflammation of the stomach and intestines. Vomiting is usually a prominent and early symptom, and purging is commonly present, the stools often betraying the nature of the poison in the shape of portions of leaves. Pain and tenderness, and distension of the abdomen are usually present, whilst giddiness and delirium are tolerably common. The post-mortem appearances are redness of the mucous membrane of the alimentary canal, the contents being sometimes discoloured. Perforation of the stomach has been seen, and occasionally peritonitis is found.

Savin.

Savin is perhaps the most important of the above, from its reputation as an abortive, most of the instances, where death has resulted from its administration, having been cases where it had been given with a view to abortion. Christison states that it does not produce abortion, in a woman not prone to miscarry, unless given in such a dose as would be likely to prove dangerous from its effects.

VEGETABLE AND ANIMAL IRRITANTS.

on the alimentary canal. Savin is usually given in the form of a powder, or a tincture prepared from the leaves or tops. The leaves, when fresh, have a peculiar odour which is easily recognised. Besides the symptoms referable to the digestive tract, savin produces hæmorrhage from the genito-urinary organs. When taken in the form of powder, this might be found adhering to the mucous membrane of the stomach, and be recognised by its odour and by its microscopical characters. The powder is soluble in water, the solution giving a deep green colour with perchloride of iron. The oil might be recognised by its odour. There are no chemical tests by which it could be detected with certainty.

Croton oil is expressed from the seeds of croton tiglium, both seeds and oil being considered very poisonous, but very few cases of death, from the administration of either, have been recorded. It is a most active purgative medicine. In searching the contents of the stomach for it, it is recommended to make an ethereal extract, and test the blistering properties of this on the skin. This would not, however, be conclusive that the extract contained croton oil. — *Croton oil.*

Colchicum is not a very common poison, but has occasionally produced dangerous symptoms, when an overdose of some quack remedy for gout has been taken. Besides the symptoms already mentioned the pupils may be dilated, the heart's action feeble, and the breathing laboured. Colchicum owes its poisonous properties to its alkaloid colchicine. — *Colchicum.*

There are three varieties of hellebore, viz. black, white, and green. The latter used not to be regarded as poisonous, but that it does possess such properties is now undoubted. In addition to the ordinary symptoms of irritant poisoning, they are — *Hellebore. veratrine* apt to produce great prostration and collapse, and sometimes insensibility. vomiting & purging

Ergot. Ergot is a fungoid growth that attacks rye, wheat, barley, and oats. It has not often been the cause of acute poisoning, the symptoms of which are vomiting and purging, colic, delirium, and stupor, and sometimes hæmaturia and jaundice. Gangrene of the extremities is the chief symptom of chronic poisoning, but convulsions have been also noticed. The blood is deteriorated, causing a tendency to hæmorrhages.

Cantharides.

This is the name given to the dried remains of the Spanish fly. It is usually administered in the form of a powder, being of a brownish-grey colour, with bright green shining particles visible in it, but there is also a tincture, which is sometimes the cause of poisoning. Its active principle is cantharidin, possibly in conjunction with some other substance, not yet determined. It has often been the cause of dangerous symptoms through being administered on account of its supposed aphrodisiac properties, or in joke, but has not often been used with intent to murder. The fatal dose has not been accurately determined.

Symptoms. Burning pain in the mouth and throat, with a sense of constriction, appear in a very short time, then pain in the epigastrium, spreading all over the abdomen, great thirst, dysphagia, salivation and vomiting, the vomit being sometimes blood-stained. There is soon pain in the loins, with frequent desire to micturate, the urine being scanty and containing blood. The genitals are swollen and priapism is common in the male; females are very liable to abort, if pregnant. Amongst other symptoms may be mentioned the formation of blisters in the mouth, conjunctivitis, lacrimation, laboured breathing, frequent pulse and convulsions, death taking place from coma. In cases which do not rapidly prove fatal, the urinary troubles are likely to persist for some time. Some persons are much more sensitive to its effects than others.

In a rapidly fatal case, if the poison has been administered in powder, some of this will be found adhering to the mucous membrane of the stomach, and may be recognised by the shining green particles. There will be intense inflammation of the œsophagus, stomach, and intestines; false membranes are sometimes found lining the inside of the mouth and throat. The bladder and kidneys are generally highly inflamed. *Post-mortem appearances.*

It is important to get the poison out of the stomach as soon as possible, and vomiting should therefore be encouraged; the stomach-pump would have to be used with caution. Oleaginous substances have been advised, but their use is not altogether to be recommended, as they would have a tendency to take up some of the powder. *Treatment.*

The naked-eye recognition of the powder has already been mentioned. If a large quantity of the poison had been taken, there might collect on the surface of the urine some drops of a green oil, which would be readily recognised. The blistering properties, when in solution, afford the only other means of recognising this substance. *Tests.*

Poisonous Foods.

Under this head would come certain kinds of fish. As is well known, some fish are poisonous at one time, and not at another. Again, all persons are not equally sensitive to the poisonous properties of fish. Thus shell fish—*e.g.* oysters, crabs, and lobsters—sometimes cause nettle-rash. Mussels sometimes induce retching and vomiting, sometimes nettle-rash, with coryza and irritation about the eyes; sometimes too there are asthmatical symptoms, whilst at other times they induce heart failure, coma, and death. Animals sometimes prove unwholesome, by reason of the food they have taken; thus hares fed on the rhododendron chrysanthemum, and pheasants fed on the laurel, have caused symptoms of poisoning, and the milk of wild goats, which had been feeding on wild herbs, has proved injurious. Snails have proved poisonous, and it is customary for those who indulge in this luxury to submit them to a few days' fasting before eating them.

Ptomaines. Diseased or putrid meat, and diseased sausages have long been known to be unwholesome, and to expose meat in such a condition for sale renders the offender liable to be prosecuted, and, upon conviction, to be imprisoned or fined. The cause of the poisoning in these cases for a long time remained obscure, but much light has been thrown on the subject in the last few years by the discovery of a group of cadaveric alkaloids, called *ptomaines*. Of these there are several different species, some having apparently no poisonous properties. They are mostly produced during the decomposition of animal matter, some observers however have asserted that they have obtained ptomaines from the normal urine of a healthy person. Besides producing the symptoms of an irritant poison, ptomaines give rise to dilatation of the pupils, giddiness, collapse, impairment of the heart's action, and death at a comparatively early stage. Both in their chemical and physiological properties they are very similar to the vegetable alkaloids. The only test by which they can be distinguished is performed as follows: The alkaloid to be tested is converted into a sulphate, and put in a watch-glass with some ferricyanide of potassium; when a drop of the perchloride of iron is added, this will be converted into the ferrocyanide, producing the well-known Prussian blue. The vegetable alkaloids also bring about the same reaction, but not so speedily as the ptomaines do. Conditions which appear to favour the development of ptomaines, are that the substance, after a short exposure to the air, should be protected from it. This is exactly what happens in the case of bodies buried, or in the preparation of tinned foods. But it has also been shown that they may be formed in tolerably fresh substances, which have not been deprived of free communication with the air. It is probable that the poisonous properties, sometimes manifested by cheese and bread that have become mouldy, are really due to the formation of an alkaloid of somewhat the same nature.

CHAPTER XXIV.

NEUROTIC POISONS.

The poisons included in this group produce their effects mainly, or entirely, on the nervous system. The most important of these is

Opium.

Opium is the crude extract prepared from the juice of the unripe capsules of the papaver somniferum. There are several varieties, as Smyrna opium, Persian or Egyptian opium, differing a little in strength according to the relative proportion of the alkaloid which they happen to contain. There is no poison that produces so many deaths as opium; thus Winter Blyth shows that between 1876 and 1880 inclusive, 40·7 per cent. of all the deaths from poison were due to opium, or one of its compounds. And further it has been stated that of all the deaths from opium poisoning, three-fourths occur in children under five years of age.

The narcotic effects of opium are mainly due to the alkaloid morphia, which exists in combination with a vegetable acid, viz. meconic acid. The other alkaloids, contained in opium, are narcotina and codeia. Opium enters into the composition of a great many preparations. Amongst those in the British Pharmacopœia the most important are the tincture, commonly known as laudanum, the compound camphorated tincture, or paregoric elixir (containing but a very small percentage of opium), syrup of poppies, the compound ipecacuanha powder or Dover's powder, and the compound soap liniment or opodeldoc. Quack remedies containing opium are also very numerous; thus Mrs. Winslow's soothing syrup contains nearly a grain of morphia to the ounce; Godfrey's cordial is of uncertain strength; Dalby's carminative contains about two and a half drops of the tincture to the ounce. Locock's pulmonic wafers, black drop, Battley's solution (the liquor opii sedativus of the Edinburgh Pharmacopœia), chlorodyne, and nepenthe may also be mentioned.

Whether the poison be taken in the form of opium or morphia, whether it be rubbed into the skin or applied to an open sore, or administered hypodermically or by enema, matters little; the symptoms, produced by an overdose, are practically the same whatever the mode of administration may have been. Within half an hour of taking the poison (and in a much less time in the case of its hypodermic injection) there is drowsiness, soon giving way to stupor with profound insensibility; at this stage the person merely appears to be in a sound sleep, and can be roused and made to answer questions, but when left to himself at once drops off to sleep. If left alone this stupor soon passes into coma, from which the person cannot easily be aroused. The face is pale and the lips livid; the pupils early become contracted, and do not respond to light; if the pupils are dilated, in a case of opium poisoning, it is usually an indication that a fatal termination is not far distant; occasionally the pupils have been noticed to be unequal. The pulse is at first small and frequent, gradually becoming fuller and less frequent, the breathing becomes slow and stertorous, the skin is often cold and bathed in perspiration, the other secretions being

Symptoms.

usually more or less suppressed. Vomiting occasionally is present, generally it is so in cases where a large dose has been taken. Speedy vomiting will sometimes be the cause of a non-fatal termination. The odour of the breath may reveal the fact of opium poisoning. In cases approaching a fatal termination there is usually complete muscular relaxation. Sometimes a person recovers from the first effects, but dies after a while in a relapse. There is another form of opium poisoning, in which the symptoms are said to commence suddenly, and these cases generally prove rapidly fatal. In some rare cases too there has been no coma, but convulsions have made their appearance. The most rapidly fatal case, on record, is one where a man died in three-quarters of an hour after taking an ounce of laudanum; if the patient survived for twelve hours there would be good hope of his recovery. Four grains of solid opium have caused the death of an adult, and two drachms of the tincture have proved fatal; one grain of hydrochlorate of morphia has also caused death, and a much smaller quantity by hypodermic injection might be fatal. Even amongst persons in health there are probably great differences in their relative susceptibilities to the influence of opium. In diseased states this idiosyncrasy is very marked; thus a diabetic person might be able to take twenty grains of opium a day without experiencing any ill effects, whereas a patient with Bright's disease might be fatally narcotised by a dose of one grain. Infants, as is well known, are exceedingly susceptible to opium, and the greatest caution must be exercised in regard to its administration to them. An infant, four weeks old, was reported to have died with symptoms of opium poisoning, after a dose of paregoric elixir equivalent to one-ninetieth of a grain of opium.

There is nothing characteristic in the appearances after death, unless the smell of opium should happen to be detected. The vessels of the brain are generally somewhat congested, but there is nothing in this to point to opium poisoning; the blood is fluid throughout the body, the heart is sometimes contracted, the stomach and lungs may be congested. In 1869 an official report was drawn up in India of a number of fatal cases of opium poisoning, and the following is a summary of the appearances most commonly found:—Brain turgid, lungs congested, heart distended with fluid blood, liver and spleen engorged, mucous membrane of stomach either natural, or slightly and uniformly injected. *Post-mortem appearances.*

The earlier the treatment is commenced the better the chance of saving the patient. If the poison has been taken by means of hypodermic injection, and if the patient be not seen until coma is well marked, the chances of success will be small. In any case there is no time to be lost. The first thing to do is to empty the stomach by means of the stomach-pump, and to continue to wash it out until the fluid is returned quite clear, and free from odour of opium. If the patient can now be roused by any ordinary means, some coffee should be put into his stomach, and he should be made to walk up and down (in the open air if possible), until all symptoms of drowsiness have passed off. If the opium has been taken by means of hypodermic injection, or if the dose was a very large one, or a considerable time had elapsed before the commencement of treatment, the plans already detailed will not suffice to rouse the patient from his state of coma. More active measures must be adopted. Of these the most useful are the use of strong induced currents, and flicking the skin with a wet towel; flagellation of the soles of the feet and calves has sometimes been found useful. These remedies must be applied vigorously if they are to *Treatment.*

have any chance of success. When the patient has been roused by these means (that is, assuming they succeed), it must be remembered that he has been submitted to a very exhausting treatment, and, if it has been long continued, it will probably be necessary that he should be given some nourishment in the shape of hot strong beef tea. Perhaps some of the cases already alluded to, where, after temporary recovery, a relapse of the coma took place, were really cases of this nature. The hypodermic injection of atropine, as the strict physiological antidote, should not be omitted in bad cases, though very often it fails to produce much effect.

Diagnosis. There are, of course, many other causes of coma besides opium poisoning, but it is only in the absence of a history that any difficulty could arise. The most difficult of all to exclude would be *hæmorrhage into the pons Varolii*, which may be followed by coma with contracted pupils and stertorous breathing; the discovery of albuminuria, and evidence of arterial changes, would point to the likelihood of cerebral affection. Lesion of other parts of the brain, though accompanied by coma, would lack the contracted pupils, these being usually dilated under such circumstances and often unequal. Possibly, too, the limbs would be more flaccid on one side of the body than the other. *Alcoholism*, in the absence of a history, may be a source of some difficulty; the face would be flushed rather than livid, and the pupils somewhat dilated; the odour of alcohol in the breath should not be allowed too much weight, as a very small quantity of spirits would suffice to produce this. *Uræmic coma* would be recognised by the presence of certain signs of Bright's disease, and by the association of convulsions with the coma, which is not so persistent and uniform as in opium poisoning. *Diabetic coma* would be recognised by the odour of the breath, the general condition of the patient, and the fact of a distended bladder, the urine on examination being found to contain sugar. After an *epileptic fit* the coma is not of so long duration, and the pupils are dilated. Thus, in a doubtful case, it is evident that it would be desirable to draw off some urine, and examine it. The diagnosis from some of the other narcotic poisons might be impossible, but this is a matter of small moment, as the treatment would be practically the same.

There are no chemical tests for opium itself, but proof of the presence of morphia and meconic acid would be considered sufficiently conclusive. Morphia crystallises in six-sided prisms, which when pure are quite white. It is but little soluble in cold water, though it dissolves readily in hot, the solution having a faintly alkaline reaction. Salts of morphia have a bitter taste. There are two tests for morphia in solution, which, when both present, are conclusive. If a solution of iodine, dissolved in hydriodic acid, be added to even a dilute solution of morphia, a crystalline precipitate of hydriodide of morphia is formed, dark red in colour, the crystals being either free or collected in radial groups. For the other test the solution, supposed to contain morphia, should be acidified with hydrochloric acid and then evaporated to dryness; the salt thus obtained is to be dissolved in as small a quantity of water as possible. This solution is then mixed with starch paste and evaporated to dryness; after cooling, a drop of a solution of one part of iodic acid in fifteen of water is added to the dry residue, when, if any morphia be present, a blue colour will be obtained.

and possesses the advantage of being applicable for all the alkaloids, and of determining also the existence of glucosides, and of other active principles derived from plants. A full description of the process will be found in Winter Blyth's book, and in the last edition of Wharton and Stillé.

Opium-eating.
Under the head of opium-eating should be included the habitual use, or rather abuse, of opium, whether it be swallowed, smoked, or taken by hypodermic injection. Writers are not agreed as to the ultimate effects on the system of this habit, some asserting that no permanent harm is done, whilst others are equally confident that the duration of life is considerably shortened. The most prominent effects of the habitual use of opium or morphia are emaciation, a yellowish tint of skin, various dyspeptic symptoms, a stooping figure and shuffling gait, and sunken eye; pains in the bones are common, when the habit has been long continued. Morphia, when habitually taken, seems to produce more effect on the mental faculties than opium. In many instances the habit has been abruptly broken off without any ill effects.

Belladonna.

This forms one of a group of substances which, besides narcotic effects, produce delirium, with illusions of the senses. The Atropa Belladonna, or deadly nightshade, belongs to the order Solanaceæ, and owes its poisonous properties to its alkaloid atropine. All parts of the plant—the leaves, root, and berries—are poisonous. Cases of poisoning by them have usually been the result of accident or negligence.

Symptoms.
The symptoms are generally uniform, and may appear in half an hour from taking the poison, or

not for two or three hours. Heat and dryness of the mouth are first complained of, the latter being so severe as to render swallowing very difficult. Contraction of the muscles of the pharynx has sometimes been present, the mucous membrane is reddened, and the voice hoarse. Dilatation of the pupils is an early and constant symptom, often becoming extreme; there is total loss of the power of accommodation. The skin is dry, and a scarlatiniform rash may appear, so that, at this stage, the case may present some resemblance to a case of scarlet fever. The pulse is rapid, and there may be slight elevation of temperature. Vomiting is not common, but the sphincters are generally paralysed. There is also some degree of paralysis of the legs, and the gait becomes very uncertain. Delirium is almost constant, the patient laughing and talking a great deal, and having spectral illusions. This stage may pass into a muttering stupor, in which the patient gradually sinks. If the patient survives seven or eight hours, he generally recovers. The majority recover. The diseases which might be confused with it are scarlet fever, hydrophobia, and delirium tremens, the latter would perhaps cause the chief difficulty. An eighth of a grain of atropine has caused dangerous symptoms, and probably one grain would prove fatal.

The vessels of the brain are generally somewhat turgid, and the blood fluid. There may be inflammation of the mucous membrane of the stomach, and upper part of the small intestine; often the mucous membrane of the former is stained of a deep purple colour by the juice. The leaves, berries, or seeds are often discovered in the contents of the alimentary canal. The seeds are small, weighing about ninety to the grain, somewhat oval and dark-coloured, they are covered

Post-mortem appearances.

with small round projections giving them a honey-combed appearance.

Treatment. The use of the stomach-pump, and of emetics, is of course of the first importance. Animal charcoal should be given, and castor oil has been recommended. Pilocarpine is the strict physiological antidote, and should be administered by hypodermic injection.

Tests. Atropia is a white crystalline substance, readily soluble in alcohol, ether, and dilute acids, when heated on platinum it burns with a yellow smoky flame. If it is covered up with a little fuming nitric acid, dried over a water-bath, and, when cold, moistened with a drop of potassa dissolved in alcohol, a violet colour is instantly produced, which is quite characteristic. The physiological test on the iris is, however, the most simple and the most reliable. A solution of 1 in 120, when dropped into the eye, will cause the pupil of an adult to dilate in about seven minutes.

Hyoscyamus.

The seeds, root and leaves of the Hyoscyamus Niger are all poisonous, owing to the presence of an alkaloid, hyoscyamine. The seeds are somewhat oval and not so round as those of belladonna; they are covered with ridges of nipple-like projections; the root is spindle-shaped, and the leaves are sessile. The alkaloid hyoscyamine is identical with duboisine (an alkaloid obtained from Duboisia Myoporoides), and has been found efficacious in calming outbursts of maniacal excitement in insane persons. Poisoning by hyoscyamus is decidedly rare. The following case, quoted by Taylor, illustrates the general course of the symptoms: A woman swallowed by mistake an ounce and a half of tincture of hyoscyamus; in ten minutes she had a hot burning, pricking sensation in the hands, feet, and legs, she became giddy and delirious, and complained of great dryness in the throat. Shortly afterwards, in attempting to get out of bed, she found that her legs were powerless. A purplish rash appeared over the body, particularly about the neck and face, which were much swollen. Four hours after taking the poison she was almost insensible, and unable to speak; her tongue was swollen, brown, dry, and protruded with difficulty, the face was swollen and scarlet, the pupils were dilated to their utmost, and the skin was hot and dry. There was no vomiting. In three hours she passed a motion smelling strongly of the drug. She ultimately recovered, but it was a week before she regained the use of her legs. There is nothing characteristic about the post-mortem appearances, and the treatment would have to be conducted upon general principles.

Cocculus Indicus

Stramonium.

All parts of the Datura Stramonium are poisonous, especially the seeds and fruit. Stramonium has been much used in India by professional poisoners. The fruit has a strong outer prickly coat; the seeds are circular, flattened, and much larger than those of belladonna; the leaves are irregular in shape and deeply ribbed. Stramonium owes its properties to the presence of an alkaloid, which both in chemical and physiological characters has been proved to be identical with atropine. Stramonium has enjoyed considerable reputation as a sedative in spasmodic asthma. One death has been recorded from smoking stramonium cigarettes. The symptoms generally come on within half an hour after the poison has been taken. They consist of great dryness in the throat, thirst, giddiness, staggering gait, dilated pupils, impaired vision, and drowsiness. This may be followed by muttering delirium and coma, or the patient may be convulsed ; the pulse is usually slow; delirium is often an early symptom, the patient being very hilarious and clutching at imaginary objects, a symptom probably due, in great measure, to the fact that his accommodation has been paralysed. As regards the post-mortem appearances, Taylor quotes the following case from the Indian Annals of Medical Science :—A professional poisoner, who, in order to allay suspicion, had partaken of some of the poisoned food which he had prepared for others, died from the effects. The pupils were widely dilated, the body covered with dust, the fingers of both hands were firmly clenched. There was great venous congestion of the brain and membranes, slight effusion of bloody serum chiefly on the right hemisphere; about an ounce of dark fluid blood was found at the base of the skull. The bloody points on section of the brain were numerous; the ventricles contained a considerable quantity of serum; the choroid plexus was unusually full of blood.

Camphor.

Camphor is prepared by distilling the wood of Camphora Officinarum with water, and is then re-sublimed. It is not by any means a common poison, and has but very rarely proved fatal. The symptoms produced by an overdose are giddiness, delirium, impairment of vision, numbness, tingling, and coldness of the extremities, flushing of the face, frequency of the pulse, and sometimes convulsions. Its action is increased if the camphor be dissolved in alcohol. As regards the post-mortem appearances, the stomach and intestines might show some signs of inflammation, but the chief thing would probably be the odour of the camphor, which is unmistakable. Some of the camphor might be found adhering to the walls of the stomach. In organic fluids it is easily dissolved out with chloroform, and if present in an alcoholic solution may readily be obtained by adding water, when it is precipitated. It burns with a rich yellow smoky flame.

Nitro-benzine and Aniline.

It is hardly necessary to do more than mention the fact that these are poisonous. In the case of the former the symptoms are apt to come on with great suddenness, the patient becoming livid and unconscious, with dilated pupils and stiffened limbs. They both produce marked changes in the blood.

Poisonous Fungi.

According to Dr. Badham, there are over five thousand species of fungi, of which only a few may be eaten with impunity. Some are poisonous at one time of the year, and not at another. Persons vary much in their susceptibility to mushrooms, and some varieties are eaten on the Continent which are here treated as dangerous. Mushrooms owe their poisonous properties to the presence of muscarine. Woodman and Tidy quote, from Bentley, the following data as to edible and poisonous varieties: The edible kind grow solitary in dry places; they are generally white or brownish; they have a compact, brittle flesh; they do not change colour when cut and exposed to the air; their juice is watery, their odour agreeable, and their taste neither bitter, astringent, acrid, nor salt. The poisonous varieties grow, in clusters, in woods and dark damp places; they usually have a bright colour; they have a tough, soft, watery flesh; they change to a brown, green, or blue tint when cut and exposed to the air; their juice is often milky, their odour powerful and disagreeable, and their taste either bitter, astringent, acrid, or salt.

Symptoms.   Vomiting and purging, giddiness, delirium, and convulsions are amongst the chief symptoms; in some cases coma has come on very soon. If the patient survives the first twelve hours, there will be a good prospect of his recovery; the majority recover. There is nothing particular to note as to the post-mortem appearances. The blood, it is said, is unusually fluid. If portions of mushrooms were found in the stomach after death they should be scraped, and the scrapings examined under the microscope for spores, by means of which the species might be determined.

CHAPTER LXXV.

ALCOHOL, ETHER, CHLOROFORM, CARBOLIC ACID.

Alcohol.

Of the effects of chronic intemperance, delirium tremens alone concerns the medical jurist. This state renders a person not liable for his actions, but a mere condition of drunkenness carries no such privilege with it. There is probably no organ or tissue in the body which may not, sooner or later, be affected by habits of intemperance, and questions as to the effects of such habits may arise in regard to life insurance; but in such cases it is generally rather the concealment, or denial, of drunken habits, than their results on the system, that constitutes the subject in dispute.

Alcohol may act, when taken in a large quantity, as a very active poison. Thus several cases where death has occurred within half an hour have been recorded. In such the person may fall down insensible, almost whilst drinking the spirit, the face becomes bloated, the breathing stertorous, and the pupils dilated, death ensuing from gradually deepening coma. When a smaller, but still poisonous, quantity has been taken, there will be confusion of ideas, inability to stand, a flushed face, afterwards giving way to pallor, the eyes suffused, the pupils, at first contracted, later on dilated and insensible to light, the breath smelling strongly of

Symptoms of acute poisoning.

the spirit that has been taken. The pulse, from being more frequent, gradually becomes slower, and the breathing becomes stertorous. In children convulsions have been noticed to be tolerably common at this stage. One striking peculiarity, in cases of acute poisoning by alcohol, is the tendency to remission often observed in the symptoms, the patient appearing to recover entirely, and then dying suddenly; this may happen some hours, or even a few days, after the poisoning. The fatal dose depends upon the strength of the spirit taken. Taylor records the death of a child, aged seven years, from four ounces of brandy taken at a dose.

Post-mortem appearances. The stomach will be intensely congested, when death has resulted from acute alcoholic poisoning, the duodenum and the rest of the intestines may also be congested. The blood is usually dark and fluid, the sinuses of the dura mater being full of blood. The right side of the heart and veins of the head and neck will be engorged, whilst the left side of the heart is empty. Œdema of the lungs is often present, and the abdominal veins may be full. Alcohol retards putrefaction, and delays cadaveric rigidity; thus, in one instance, on the ninth day after death the body was comparatively fresh. The odour of alcohol is usually very marked, and has been relied upon as a proof of the cause of the symptoms, but it may disappear rapidly.

The history of the case will usually furnish the means of distinguishing between acute alcoholic poisoning, and concussion of the brain or opium poisoning, with which states it has many features in common. As regards treatment, the rules that are applicable in the case of opium poisoning, apply also here. The main thing is to get the poison out of the system by emetics and the stomach pump, as far as possible, and to rouse the patient by means of the battery, &c.

Treatment.

Alcohol may be obtained from any substance *Tests.* containing it by distillation. The distillate should be mixed with chloride of calcium, and the mixture shaken up with carbonate of potash; when this is allowed to stand for a little while, the alcohol will rise to the surface, and can be decanted off and tested. Alcohol evaporates without leaving any residue. The following is a delicate test, where the quantity is small. Make a mixture of strong sulphuric acid and saturated solution of bichromate of potash, dip some fibres of asbestos in this, and suspend them in the glass tube which leads from the retort in which the distillation is taking place; the asbestos gradually becomes green, owing to the formation of oxide of chromium, if any alcohol be present.

Ether.

Ether is but rarely taken by the mouth in a poisonous dose; the effect would be the sensation of heat and burning in the mouth and throat, followed by symptoms of intoxication. People have been known to use ether as a stimulant, but such ether drinking is uncommon. The vapour of ether is, however, used very extensively in this country, and almost exclusively in the United States for its anæsthetic effects. Compared with chloroform, it *Ether versus* may be noted that a rather longer time is required *Chloroform.* to get a person under the influence of ether, and that it is more irritating to the respiratory passages, that it does not exercise so depressing an influence upon the heart, that its administration is more likely to be followed by vomiting, and that it may set up an attack of bronchitis. Its use is generally considered to be attended with especial danger, where disease of the kidneys is present.

Ether may be recognised by its smell, by its *Tests.*

low boiling point, by its burning with a yellow smoky flame, and by its action on chromic acid, the test, which was described under the head of alcohol, would, if repeated, produce a precisely similar result.

Chloroform.

Chloroform is best known as an anæsthetic, but it has caused death on several occasions, through being taken in the liquid state. Immediately on swallowing the fluid there is a burning in the mouth followed by numbness. Then excitement and delirium may supervene, followed by coma and lividity, or there may be convulsions. Jaundice has been noticed as a sequel. The post-mortem appearances, in death from chloroform thus taken, would be those chiefly of inflammation of the œsophagus and stomach.

Anæsthetic effects. The chief importance of chloroform, however, is owing to the anæsthetic properties of its vapour. Unfortunately, from one cause or another, a great many lives have been lost through its administration. *Causes of death.* In the first place, a person may die from suffocation, the vapour being presented to him in such a concentrated form that he is unable to breathe. Such cases were probably far from uncommon at first, but they never occur now. The most common cause of death from chloroform is undoubtedly heart-failure, owing to fatty or other degeneration of the walls. It is, therefore, always necessary that the condition of the heart should, as far as possible, be accurately determined before commencing the administration; particular inquiry should be made as to the existence of a tendency to faintness or giddiness, or as to shortness of breath, or palpitation on exertion. It must not be forgotten, however, that people used to die on the table during operations, before the introduction of anæsthetics at all, and that, as formerly, so now, death may be due to shock, and not to the chloroform. This fact seems to have been lost sight of too much, and if the person dies on the table the anæsthetic gets all the blame. Occasionally it happens, in trials for robbery, rape, and the like, that the prosecuting party alleges that he or she was rendered insensible by the administration of chloroform. If a woman says that the accused person held a handkerchief to her mouth and nose, and that she immediately became insensible, it is tolerably certain that she is not speaking the truth; it is possible that she might have fainted under such circumstances, but there is no drug that could have rendered her insensible in so short a time. All charges made under such circumstances should be regarded with great suspicion. It has been doubted whether it would be possible to administer chloroform to persons during sleep, so that they should come under its influence without being aroused. This certainly is possible, especially in the case of children; but probably it could only be done successfully by one accustomed to give anæsthetics, as it would be essential that the administration should be very gradual. The mode of administration, and the treatment to be adopted, will be found in the text-books.

The post-mortem appearances in a person who has died from the effects of chloroform vapour *may offer no clue* as to the cause of death. If the death have been caused in the way first described there will be the usual signs of so-called asphyxia, viz.: an engorged venous system and right heart, the left cavities being nearly or quite empty. In other cases the heart may be found empty or not; its muscular substance ought always to be examined under the microscope, as fatty degeneration is fre-

<small>Post-mortem appearances.</small>

quently found in these cases. The odour of chloroform would be the most reliable evidence.

Tests. Substances suspected to contain chloroform should be reduced to a pulp and distilled over a water bath. The distillate may be tested as follows: A little aniline and an alcoholic solution of soda lye are added; on warming, a peculiar penetrating odour is given out. This is not, however, conclusive of chloroform, as chloral and one or two other compounds give the same smell when thus treated. Chloroform reduces Fehling's alkaline copper solution when applied to a distillate. The suspected substance, which should be rendered neutral, is distilled in the presence of hydrogen, and the vapour arising therefrom is passed through a tube heated red hot at one point; at this spot the chloroform will be decomposed, and hydrochloric acid formed, which can be readily recognised by any of the ordinary tests. Or if the substance containing chloroform be mixed with a little thymol and potash, a reddish violet colour is developed, becoming more marked under the influence of heat.

Hydrate of Chloral.

This drug is a very valuable hypnotic, but, unfortunately, a considerable number of deaths have occurred from its incautious use in excessive *Symptoms.* doses. An ordinary dose produces an apparently natural sleep. When a dangerous dose has been taken the sleep is profound, the breathing stertorous, and the face is flushed, so that the person looks like a man under the influence of alcohol. At first he can be roused, but gradually the coma becomes deeper, the body temperature sinks, and he dies with or without symptoms of collapse. The fatal dose is a matter of some uncertainty, as the effects of the drug are much more readily manifested in some, than in others. A drachm would be a full dose, and anything more than this might give rise to alarming symptoms. The elimination of chloral takes place slowly in heavy drinkers, and hence additional caution is necessary in its administration to people of intemperate habits. Chronic poisoning is far from uncommon, and not a few of the inmates of asylums owe their loss of mental power to the abuse of chloral; cases both of mania and melancholia have been attributed to its effects. Amongst other results of its habitual use may be mentioned an erythematous or scarlatinal rash, often followed by desquamation and sometimes by albuminuria; sometimes an acute eczema is so produced. The system does not become habituated to its use as it does to the other narcotics, and consequently death has sometimes resulted, when a dose in excess of the usual one has been taken. The post-mortem appearances are those common in asphyxia; the vessels of the brain are engorged, the subarachnoid space and the ventricles contain an excess of cerebro-spinal fluid; the brain itself may appear somewhat shrunken. The mucous membrane of the larynx is injected and somewhat œdematous. The lungs are congested, the right side of the heart being distended and the left empty. *Post-mortem appearances.*

One of the main objects of treatment is to maintain, or rather restore, the body heat. This is effected by friction to the skin, by placing the patient in a warm atmosphere, and by causing him to inhale warm air. Artificial respiration usually has to be performed, and the use of the battery currents (both constant and induced) will often be of assistance. As regards the hypodermic injection of strychnia, authorities are by no means agreed; some say that it is the best antidote, others that it is unreliable. *Treatment.*

Hydrate of chloral is decomposed by the addition of an alkali, such as soda or potash, into chloroform, water, and formiate of soda or potash. It is soluble in one and a half time its weight of water; the solution should be neutral. There should be no cloudiness when a solution is tested with nitrate of silver in the cold; if, however, the solution of nitrate of silver is boiled and a little ammonia added, there will be formed a mirror of reduced silver.

Bisulphide of Carbon.

Carbon bisulphide is a colourless fluid, which, unless pure, has a penetrating and very disagreeable odour. It is a good deal used for dissolving phosphorus, caoutchouc, gutta-percha, &c. A case has been recorded, in which a man took two ounces with suicidal intent. The chief symptoms were pallor, dilated pupils, a frequent pulse, lowering of the body temperature, and convulsions. He ultimately recovered. It is so apt to produce ill-health, amongst those who are constantly using it in their occupation, that it possesses an interest for the medical jurist. There are two periods or stages in those suffering from its effects, marked by different symptoms. The first period—that of excitement —is characterised by headache and digestive disturbances, curious creeping sensations are perceived in the skin, and there is a great change of temper, the patient becoming irritable, and suffering from tinnitus; in one instance the patient became maniacal. The other period is marked by a general depression. There is anæsthesia of the skin, and not of the skin only, but also of the mucous membranes, patients complaining of a feeling in their tongues as if they had been tied with a cloth. The limbs, too, become paralysed, and quite recently it has been shown that some patients become amblyopic, one patient having also been noticed to be completely colour-blind. As to the permanency, or otherwise, of these symptoms nothing at present is known. Carbon bisulphide is recognised by its odour, by its boiling-point, which is 47°, by its action on an alcoholic solution of potash, xanthogenate of potash being formed, which gives a yellow precipitate with sulphate of copper, and, lastly, by the black precipitate of sulphide of lead, which is obtained when it is heated with nitrate of lead and potash.

Carbolic Acid.

This is obtained in a solid crystalline form by distilling coal tar. It is used very extensively as a disinfectant and antiseptic. Poisoning by means of it has become very common during the last few years, so much so that it now stands sixth in the list of fatal poisons in England. Death from its use is generally the result of accident, occasionally it is suicidal; its strong and disagreeable odour would almost preclude its administration homicidally, except perhaps in the case of children.

Symptoms.—The poison is readily absorbed when applied to a wound, and produces exactly the same symptoms as when taken internally; it also acts through the unbroken skin, or mucous membrane, producing in addition a local numbing effect. It is readily absorbed when administered in the form of an enema. When taken internally, there is a burning sensation in the mouth with a very disagreeable taste; then vomiting, a feeling of faintness with cold perspiration, and a small pulse; the pupils become contracted and insensible to light, and the patient becomes insensible; the breathing gradually becomes shallower, and the

T

patient dies from failure of respiration, often with marked depression of temperature. Convulsive twitching of groups of muscles is a not uncommon sign, sometimes only seen in the face, at others only in the limbs. The condition of the urine is one of the most reliable guides in practice as to the extent to which carbolic acid has been absorbed.

Carboluria. In a well-marked case of carboluria, the urine will be of a deep olive-green colour, almost black, and the odour of carbolic acid may be distinctly perceptible; it will keep for many days without any sign of decomposition, and becomes darker in colour by keeping. Carbolic acid is eliminated in the urine as phenylsulphate of potash, having combined with the sulphates normally present in the urine. When the sulphates have all been used up in this way, toxic symptoms will begin to appear. It is desirable, therefore, when carbolic acid is being used as a dressing, that the urine should be occasionally tested for sulphates, for so long as these are present there is no danger of carbolic acid poisoning. This is easily done by adding to the urine acetic acid and some chloride of barium, when a precipitate of sulphate of barium is thrown down; when this precipitate can no longer be obtained poisoning is imminent, and it is said that it may be averted by the internal administration of an alkaline sulphate, such as sulphate of soda, which may be given in the form of a mineral water. The course of carbolic acid poisoning is very rapid; in some cases collected by Falck, one-third died in the first hour, 71 per cent. died within twelve hours, and 91 per cent. within twenty-four hours. About 230 grains would probably be a fatal dose, though a case of recovery has been recorded after ten times that quantity had been taken.

Post-mortem appearances. There will be brownish stains about the lips and mouth, the mucous membrane of which, as well as that of the throat and œsophagus, will be white, sodden, and sometimes eroded. The mucous membrane of the stomach may be corrugated and thickened, or sometimes it is altogether destroyed. The small intestines may be affected in a similar way to a less degree. The bronchi are inflamed, and the lungs somewhat œdematous. The vessels of the brain, the right side of the heart, the kidneys, liver, and spleen are generally congested. Putrefaction is much delayed, and, on opening the body, the odour of carbolic acid will be very plainly perceived.

The stomach should be thoroughly washed out with tepid water until the fluid no longer smells of the acid; this will in practice be found rather difficult to attain, but the washing should at any rate be continued, until the fluid returned is free from colour. If the coma is deep, the battery will have to be used as in a case of opium poisoning. Olive oil has been recommended as an antidote, as also has a saccharated solution of lime given in an excess of water, but, as this requires three days for its preparation, it is not likely to come into very general use. *Treatment.*

Chips of pine wood give a beautiful blue colour when dipped first in carbolic and then in hydrochloric acids, and exposed to the light. As, however, some species of pine give a blue colour with hydrochloric acid alone, it is best to test them separately with this first. If to a solution containing carbolic acid ammonia be added, and then a small quantity of a solution of hypochlorite of soda, a blue colour appears; this reaction is hastened by heat. Perchloride of iron gives a violet colour with carbolic acid. With a neutral or acid solution, bromine water gives a precipitate of very fine stars of needles; this is a very delicate reaction. *Tests.*

CHAPTER XXVI.

STRYCHNIA, CONIUM, CURARE, PHYSOSTIGMA, TOBACCO.

Strychnia.

THIS is the alkaloid obtained from the seeds of *Strychnos nux vomica*. The seeds also contain brucine, igasurine, and strychnic or igasuric acid. The seeds are round, flattened, of a greyish-brown colour, and rather smaller than a broad bean; the surface of the seed is covered with very fine silky hairs, which can be recognised under the microscope when the powder obtained by crushing the seeds is examined. This powder is of a greyish-brown colour, of extremely bitter taste, and is easily recognised by the deep orange-red which it gives with nitric acid: this is not due to the strychnine, but to the brucine contained therein, and is conclusive of the presence of nux vomica in the powder. Nux vomica forms the chief ingredient of many vermin-killers and rat-killers, and thus is readily accessible to the public. The poisonous symptoms of nux vomica are practically those of strychnia, and do not therefore require a separate description.

When swallowed there is immediately noticed an intensely bitter taste. The symptoms commence with great suddenness, usually within twenty minutes of the poison being taken. The person feels as if he could not get his breath, and wants fresh air; he complains that he is going to be choked, and is very restless. Twitchings and jerkings in the muscles set in, followed by violent tetanic spasms, attacking almost all the muscles of the body at once. Opisthotonos is generally present, so that, in an extreme case, the body rests on the head and heels, the arms are rigidly extended by the side, and the feet are arched; the abdominal muscles are tense, and the chest fixed, so that respiration is stopped, and the patient is in imminent danger of being suffocated. The face and lips become livid, and the eyeballs staring; sometimes the corners of the mouth are drawn up by the muscular contraction, producing the risus sardonicus. The patient suffers much from thirst and dryness of the throat. The mind is unaffected, and, in the intervals of the paroxysms, the patient often foretells his death. He can usually tell when a paroxysm is coming on, and screams out, asking to be held; sometimes during the attack there is foaming at the mouth. The jaw is not affected as a rule until late, and during the intervals is perfectly flaccid, a point of great diagnostic importance. The pupils are said to dilate during the paroxysm, and contract in the interval. In cases that prove fatal the severity and frequency of the attacks keep on increasing to the last. The duration of the paroxysm may vary from half a minute to eight minutes; the average duration is perhaps about two minutes. One-fifteenth of a grain killed a child aged three; two grains might be regarded as a fatal dose for an adult. In the most rapidly fatal case on record death took place in five minutes; it is usually considered that if a person survives for two hours there is fair hope, though by no means any certainty, of his recovery. Patients often derive relief from being rubbed or held during a paroxysm.

The murder of John Parsons Cook by William

William Palmer. Palmer, by the administration of strychnia, led to the first trial for alkaloidal poisoning in this country. Dr. Taylor's full report will be found in the Guy's Hospital Reports. Briefly, the facts were as follows: Cook had been staying with Palmer at Shrewsbury, and was there taken ill, vomiting and abdominal pain being the chief symptoms; there was some reason afterwards to think that he might have been at this time suffering from antimonial poisoning. They removed to Rugeley, where Palmer practised, for he was a medical man, and a neighbouring practitioner was called in to attend Cook. Amongst other things some pills were prescribed for him to take at night, containing calomel and morphia. At first he went on well, but on the third night, about an hour after taking the pills, he screamed out twice, and said that he should die; his head was in motion, jerking backwards, his arms were straightened out, and his legs set quite stiff, the eyes were staring, the head was drawn back, and the mouth closed; he rallied after some hours, and by next morning seemed to have quite recovered. The next evening he did not want to take his pills, but Palmer and the other practitioner explained their nature to him, and he gave way. An hour after taking them he was seized with convulsions, and a feeling of suffocation. A medical man, a friend who had only come to see him that day, thus described his state in his evidence:—'I have never seen convulsions so strong before. They were symptoms of tetanus; every muscle in his body was stiffened.' His head and neck were affected with spasms, his head was thrown back, his hands were clenched, and his arms were in a state of rigidity, his jaw was fixed and closed, his body rested on the head and heels, and he died in less than twenty minutes. Owing to circumstances which need not be detailed here, the post-mortem examination yielded negative results. The coroner's jury nevertheless returned a verdict of wilful murder against Palmer, and he was subsequently tried and executed. 'The attacks above described,' says Taylor, '' on two successive evenings were of exactly similar character, coming on each time about an hour after the taking of some pills. He had recovered on the first evening, and remained comparatively well all the following day, and then, without any assignable cause, he died with symptoms of tetanus in less than twenty minutes.' Such symptoms do not belong to the course of any known or conceivable disease, and the theory of the prosecution was, that Palmer had substituted strychnine pills for those prescribed by the practitioner in attendance.

If death took place during a paroxysm, the body would probably retain the attitude it had at the moment of death. Otherwise the body would be relaxed at the time of death. Rigor mortis sets in early as a rule, and lasts a long time. Thus in some cases—where, for instance, the body was in a state of opisthotonos—the external appearances might be such as to suggest strychnia poisoning. The brain and its membranes, and the upper part of the spinal cord, usually show some congestion. The lungs are congested, the heart empty and contracted, or the right side distended, the blood fluid and dark. The mucous membrane of the stomach may be congested; the bladder is generally empty. *Post-mortem appearances.*

The stomach should be thoroughly washed out by means of the stomach-pump; it is sometimes necessary, owing to the closure of the jaws, to administer chloroform before this can be done. By way of medicinal treatment, chloral hydrate, Calabar bean, and bromide of potassium have been much lauded; the first is probably the most useful in subduing the convulsions. Tracheotomy may sometimes be necessary. *Treatment.*

Diagnosis. The disease with which poisoning by strychnia is most likely to be confounded is tetanus, especially when idiopathic. In tetanus the jaw is always first attacked, and remains stiff all through the disease, while in strychnia poisoning the jaw is not attacked until late in the disease, and during the intervals between the paroxysms is usually quite free from rigidity. If it could be proved beyond doubt that the person had not taken any food, or had an enema, or a hypodermic injection of anything, or any local application to the skin, for a period of several hours, then the case could not be one of strychnia poisoning; on the other hand, if the patient died within two or three hours of the commencement of his illness, and if that illness came on within twenty minutes or so of his taking some food, then the symptoms almost certainly might be attributed to poisoning. Cases of poisoning by strychnia have sometimes been mistaken for hysteria; thus on one occasion a young lady was certified to have died of hysteria, and it was afterwards proved, by the confession of her murderer, that she had died from the effects of strychnia. Such cases are less likely to occur now, but the difficulty is sometimes a very real one. Occasionally, too, in pregnant women there has been some confusion with puerperal convulsions, but in these there is seldom a stage of complete remission between the seizures, and death rarely ensues before the birth of the child.

Tests. Strychnia is exceedingly bitter. If a neutral solution of chromate of potash is added to a solution containing strychnia, a crystalline chromate of strychnia is formed, in the shape of orange-yellow needles. If these crystals are put on a piece of white porcelain, and a drop of sulphuric acid added, a deep rich blue colour rapidly appears, passing through purple into red. Similar colours may be obtained by adding sulphuric acid to strychnia in the presence of ferricyanide of potash, permanganate of potash, peroxide of lead, or peroxide of manganese. In the subliming cell strychnia sublimes at 169° in the form of minute needles, consisting of lines, mostly straight, with parallel feathery lines at right angles. Letheby's galvanic colour test is performed as follows: the residue, obtained by evaporating a strychnia solution in a platinum dish, is touched with a drop of concentrated sulphuric acid, the platinum is connected with the positive pole of a Smee's battery, the negative pole being brought into contact with the acid; when the current passes a violet colour flashes out, which remains after the pole is removed. The physiological test consists in injecting under the skin of a frog some of the solution suspected to contain strychnia, and comparing the effects with the known effects of strychnia.

Conium.

Conium maculatum, or hemlock, belongs to the order Umbelliferæ. All parts of the plant—seeds, root, and leaves—are poisonous. The poisonous properties are due to its alkaloid conia. The great majority of cases of poisoning by hemlock are the result of accident, the leaves being often mistaken for those of parsley. A feeling of languor and loss of muscular power are amongst the earliest symptoms, the latter gradually passing into actual paralysis, the legs feeling numb. In the early stages dryness of the throat is complained of; the pupils are at first contracted, later on they become dilated, and paralysis of accommodation is noted. Death may take place from heart failure, or from stupor gradually deepening into coma. There is

Symptoms.

Post-mortem appearances.

nothing characteristic after death; the membranes of the brain are often congested, and the cerebro-spinal fluid increased in quantity. The lungs are engorged. When the leaves have been taken a green pulpy mass will be found in the stomach, from an examination of which the nature of the poison might be ascertained.

Treatment.

The treatment consists in giving emetics and washing out the stomach; the earlier it is commenced the better the chance of saving the patient.

Tests.

The plant has a disagreeable odour when bruised, and when any part of it is rubbed up with caustic potash, an odour is brought out, which is compared by some to the odour of white mice. Conia may be recognised by having the same unpleasant odour as the leaves, when treated with potash. Sulphuric acid, added to bichromate of potash in the presence of conia, causes the formation of butyric acid, which may be known at once by its odour.

Œnanthe Crocata.

This, the hemlock water-dropwort, also belongs to the order Umbelliferæ. The plant has some resemblance to celery. The stem is round, channelled, smooth, branched, and of a yellowish red colour. The root consists of a series of oblong tubercles with long, slender fibres, and is the most poisonous part of the plant. The leaves are of a dark green colour with a reddish border. It is one of the most deadly poisons, though the nature of its active principle is not known. A number of convicts at Woolwich ate some of the leaves and roots. In about twenty minutes one man was seized with convulsions, his face became bloated and livid, his breathing stertorous, and he died in five minutes. A second man died in a quarter of an hour with similar symptoms, although the stomach-pump was used and some of the leaves were extracted. A third died in about an hour, and a fourth not long afterwards, in spite of most energetic treatment by cold affusion, emetics, the use of the stomach-pump, stimulants, and stimulating frictions. Two other cases proved fatal, one in nine, the other in eleven days. In these there was found irritation of the alimentary canal; in the other cases there was nothing very definite made out; the blood was fluid and very dark, the lungs were engorged, the stomach and intestines were externally of a pink colour, internally the stomach was lined with a thick viscid mucus containing portions of the root. In another instance, eight boys ate some of the plant, and five of them died within twenty-four hours. There are no special tests; recognition of the plant would be based upon its botanical characters.

Cicuta Virosa.

The water-hemlock also belongs to the Umbelliferæ. The umbels are large, the leaves tripartite, the leaflets linear-lanceolate, acute, serrate, decurrent. All parts of the plant are poisonous. An active principle named cicutoxin has been separated from the root, but not much is known about it at present. The symptoms to which it gives rise are pain and burning in the stomach, headache and vomiting, and convulsions. The breathing is stertorous, the face cyanosed, the pupils dilated, the pulse small, and death may supervene in the course of a few hours. There are no known reliable tests for it.

Æthusa Cynapium.

This is commonly known as the fool's-parsley. It may be known from parsley by the fact that the leaves are finer, more acute, and of a darker green colour than those of parsley. When rubbed it gives out an odour quite distinct from that of parsley. Dr. John Harley, who has performed many experiments on himself, and others, with the juice of the plant and with tinctures prepared from the ripe and green fruit, believes that the plant is quite harmless, and that in cases where dangerous symptoms have been attributed to its use, there must have been some mistake, conium or aconite and not æthusa having been taken.

Curare.

This is obtained from a South American plant called curari, and is used by the Indians as an arrow poison. The juice is extracted from the stem, and then mixed with certain other ingredients, such as red and black ants, the whole mass being heated. It acts by paralysing both the voluntary and respiratory muscles, the heart being unaffected. It has been shown that it does not affect the nerve centres, but the intra-muscular terminations of the motor nerves. It has been used with good effect in hydrophobia and tetanus by hypodermic injection, as its effects are not manifested when it is taken by the mouth. The active principle is curarine, which gives an amorphous precipitate with bichromate of potash, the precipitate becoming blue on the addition of sulphuric acid. Curarine is directly antagonistic to strychnia. In the rare cases where poisonous symptoms have appeared from its use, artificial respiration has been necessary until all symptoms have passed off. Strong coffee might be given, and free perspiration induced.

Taxus Baccata.

The yew is a tolerably common poison, the leaves or berries being not unfrequently eaten.

Contrasted with the savin leaf, the yew leaf is not so pointed, and lacks the peculiar odour of savin. Yew berries are of a light red colour, about the size of a pea, open at the top, and contain a large ovoid brown kernel. The symptoms are pallor, a small pulse, cold extremities, dilated pupils, convulsions, insensibility, and coma. A table-spoonful of the fresh leaves was administered to three children of five, four, and three years of age; the eldest vomited a little and complained of pain in the abdomen; the others suffered no pain, they were simply listless. They all died in a few hours. The alkaloid taxine gives a red colour with sulphuric acid.

Physostigma Faba.

The Calabar bean, or *Physostigma venenosum*, comes from the West Coast of Africa, where it is used as an ordeal bean, when a person is suspected of witchcraft; the popular superstition being that innocent persons, who take it, vomit and are safe, whilst the guilty retain the poison and die from its effects. The bean has a hard brown covering, and is kidney-shaped, measuring an inch in length by half an inch across; along the back of it there is a furrow with raised edges, and at one end there is a small hole. It owes its poisonous properties to an alkaloid, physostigmine, better known in the medical world as eserine; it also contains a second alkaloid, calabarine. The late Sir R. Christison experimented on himself with the drug when it was first brought over to this country. Having taken six grains overnight without much effect, the next morning he took twelve grains, and very soon suffered from giddiness, weakness, and faintness. His pulse became very feeble and irregular, and he remained much collapsed for two hours, after which he went to sleep, and by next day all the symptoms had disappeared. Some years ago a wholesale poisoning by Calabar bean occurred at Liverpool. Some of the beans had been emptied out of a ship's hold on to a rubbish heap, where they were found by a number of children, who eat a quantity. Within an hour many of them showed symptoms of illness, and no less than forty-six were taken to the hospital. In all there was some shock, amounting in several to collapse; in thirty-eight there was vomiting, and in eighteen diarrhœa. One child only—a little boy aged six—died; he had no vomiting, but was much collapsed, and died from syncope. At the post-mortem examination nothing very definite was found.

Eserine, when applied locally to the eye, causes a marked contraction of the pupil in about ten minutes; when taken internally it also produces contraction of the pupil, but not so markedly. In large doses it paralyses the vagus, and also acts on the intra-muscular terminations of the nerves, at first exciting and afterwards paralysing them. It has been used subcutaneously in cases of tetanus to control the spasms. In case of an overdose atropine should be injected.

The following is an exceedingly delicate test for the alkaloid: A solution of bromine in water, when added to a solution of sulphate of eserine, produces a red colour, when less than the thousandth part of a grain is present. Concentrated sulphuric acid dissolves eserine, producing a yellow colour, changing to olive-green; ammonia and the carbonated alkalies precipitate the alkaloid, from an acetic acid solution, in the form of oily drops. When dropped into the eye of a rabbit it produces contraction of the pupil. A solution applied to the web of a frog's foot causes dilatation of the blood-vessels, and, if injected subcutaneously, it produces paralysis of the respiratory and voluntary muscles, and death.

Tobacco.

Tobacco is prepared from the leaves of *Nicotiana tabacum*, and has occasionally been the cause of death when smoked, swallowed, or taken in the form of an infusion. The general symptoms of poisoning by tobacco are nausea, vomiting, colic, diarrhœa, frequent micturition, giddiness, faintness, pallor, trembling, a small, weak, and irregular pulse, tetanic cramps, and great collapse. Thirty grains, in the form of infusion, have proved fatal. There are no special post-mortem appearances known; the blood is generally fluid. The treatment would be conducted upon general principles.

The effects of tobacco smoking do not differ from the above except in degree, and it is astonishing how soon the system becomes habituated to its use. Whether moderate smoking does harm need not now be considered, but there can be no question that excessive smoking is prejudicial. Many cases of disordered digestion, and of palpitation of the heart, can be traced to it, and there is a special form of amaurosis, the result of chronic inflammation of the optic nerve, ending in recovery or partial atrophy, which is said to be due to nothing but tobacco. Of course the kind of tobacco that is used makes a difference, and the amaurosis has hitherto been found to follow the use of strong tobacco in excess. Whether any other disorders of the nervous system are caused by it, or not, there is not enough evidence at present to show. The symptoms in all cases of tobacco poisoning are due to nicotine, which appears to be one of the most potent poisons known. Hitherto only three cases of poisoning by it have been recorded—two of

[marginal note: Tobacco smoking.]

suicide and one of murder. The latter is worth mentioning, as it was the first occasion on which an alkaloid was used as a poison.

Poisoning by nicotine. In 1851 the Count and Countess Bocarmé were tried for the murder of her brother with nicotine, and convicted; the poison had been forcibly administered, and the victim died within five minutes. It appeared that the Count had been studying chemistry for some months, especially as to the extraction of nicotine from tobacco, and, after the murder, he tried to conceal the crime by pouring strong acetic acid into the mouth, and over the body of the deceased. The viscera and some scrapings from a wooden plank, on to which the deceased had vomited, were examined by Stas, and found to contain a very large quantity of nicotine.

Nicotine first excites and then diminishes the activity of the brain and spinal cord; it first quickens and then slows, and finally stops the respiratory movements, whilst as regards the heart it causes it to beat fast, irregularly, and weakly; the blood-vessels are at first narrowed and then dilated. Chlorine gas colours nicotine blood-red or brown; chloride of platinum throws down a reddish crystalline precipitate. Nicotine resembles ammonia in several of its reactions, but the odour ought always to distinguish them.

Cocculus Indicus.

The Levant nut, as it is commonly called, is the fruit or berry of the *Anamirta Cocculus*, imported from the East Indies. The symptoms, that the powdered seeds give rise to, are nausea, vomiting, and griping pains, followed by stupor and intoxication. The symptoms are due to the presence of an alkaloid, called picrotoxin. It has been used very frequently for the adulteration of beer; a solution of it is also used a good deal in the United States for destroying lice, and three deaths have been recorded from its being taken in this way by mistake.

The following tests are recommended for picrotoxin when pure: in cold concentrated sulphuric acid it dissolves, forming a golden yellow colour, which becomes violet on the addition of bichromate of potash. If picrotoxin, tolerably pure, is dried and mixed with three times its bulk of saltpetre, and the mixture moistened with sulphuric acid, and then decomposed with soda-lye in excess, there is produced a transitory brick-red colour. This is a very delicate test.

Lobelia.

The *Lobelia inflata*, or Indian tobacco, is much employed by quacks. The seeds and leaves are both used, and owe their properties to the presence in them of lobelin. Lobelia is used extensively in America by a class of practitioners, who call themselves Thomsonians. The symptoms it produces, in an overdose, are those of a powerful emetic, nausea and vomiting, with pain at the pit of the stomach, being early and constant symptoms. Difficulty in breathing, speaking, and swallowing follow, the pulse becomes very feeble, and there is great muscular weakness. The powder, obtained by pounding up the leaves, gives a reddish-brown colour with nitric acid, and is blackened by sulphuric acid. The seeds of lobelia are very small, oval in shape, with a reticulated surface and hairs projecting from it.

CHAPTER XXVII.

PRUSSIC ACID, DIGITALIS, ACONITE.

Prussic Acid.

PRUSSIC acid, or hydrocyanic acid, is obtained from several plants, amongst them being the cherry-laurel, and bitter almond plant. It is also obtained from the kernels of the peach and apricot, and from the roots of the bitter and sweet cassava. Pure prussic acid is exceedingly poisonous, and it and its compounds are frequently used by suicides, so that, in this respect, it ranks next to opium, 35 per cent. of the suicides from poison, in this country, being due to the cyanides in one form or another. The acid is very volatile, and its vapour is of course highly poisonous. There are many ways of taking or administering prussic acid, without using the acid by itself; thus poisonous symptoms have been known to follow the eating of a quantity of bitter almonds, or of peach kernels, and death has followed the taking of the oil obtained by distilling the pulp of bitter almonds. Bitter almond water, prepared by distilling almond cake in water, is very poisonous; laurel water, prepared from laurel leaves, is also an active poison. Cyanide of potassium is quite as often used as the pure or dilute acid, and is quite as poisonous. As in all these cases the active principle is the same, one general description of the symptoms will suffice for all.

Various modes of administration.

Sometimes the patient becomes insensible so rapidly, that it is impossible to observe any symptoms. If the case should not prove so rapidly fatal, there will generally be a feeling of constriction in the throat, nausea, giddiness, and sometimes convulsions; trismus is not uncommon. The person then falls to the ground insensible; his eyes are fixed and glassy, the pupils moderately dilated, and not acting to light; the muscles relaxed; the skin cold and clammy, and somewhat livid; the breathing is peculiar, expiration being much prolonged; sometimes it is stertorous. The pulse very soon becomes imperceptible. The odour of prussic acid, which is quite characteristic, and has been likened to that of the peach blossom, will generally be recognised about the mouth of the patient. The fæces and urine are often passed involuntarily, during the stage of insensibility. Death generally takes place, when a large dose has been swallowed, in ten minutes or thereabouts, sometimes as quickly as two minutes. The longer the patient lives after an hour, the greater will be the chance of his recovery. Probably one grain of the pure acid would be fatal. The question has arisen in the law courts, on a good many occasions, as to what voluntary acts a person could perform, after taking a fatal dose of prussic acid, some authorities doubting whether even such simple acts as putting a cork in a bottle, and putting it under the mattress, were possible; but the following case, mentioned by Taylor, ought to set the matter at rest. A gentleman swallowed a quantity of prussic acid; he then walked ten paces, descended a flight of seventeen steps, and walked forty paces further, entering the druggist's shop where he had previously bought the poison, and said, 'I want some more of that prussic acid.' He then became insensible, and died in the course of about ten minutes. In

death usually preceded by a shriek

some cases the person utters a cry at the moment of becoming insensible, but the absence of this, or of convulsions, could not be allowed to have the least weight, as evidence against the theory of prussic-acid poisoning. There is no disease which could be confounded with prussic-acid poisoning; rapidly fatal cases of syncope or apoplexy would always be distinguished, without difficulty, on the post-mortem table. As regards opium poisoning, the history of the case would prevent any confusion; in the absence of any history, the chemical tests would supply the distinguishing evidence.

Post-mortem appearances. Cadaveric rigidity comes on early in these cases. The conditions noted as present during life will still be found: the jaws will be fixed, the eyes glassy, the pupils dilated, the nails livid, the hands and feet contracted. Unless the body has been exposed to the air, there will probably be an odour of the poison about it. If this should not be present, it would be well to follow Casper's advice, and open the head first, as the brain undergoes putrefaction less quickly than the contents of the thorax and abdomen, and therefore would be likely to retain the odour longer. The absence of the odour, however, cannot be admitted as evidence against poisoning, as, in many undoubted cases, it has not been detected. On opening the body, the brain may or may not be congested, the lungs and right side of the heart are generally engorged, the lungs are often œdematous, and the bronchi contain some frothy mucus. The blood is generally dark and liquid, sometimes it is thick; prussic acid combines with the hæmoglobin, and it has been shown experimentally that blood, when thus treated, will not absorb oxygen. The stomach and intestines do not present any constant lesions.

Treatment. Seeing the rapidity with which the changes come on, and the great alteration that takes place in the character of the blood, it can readily be understood that chemical antidotes are not of much avail. The stomach-pump might be used, and the contents of the stomach thoroughly evacuated, or an emetic might be given, if the patient could swallow. Unless the patient is seen early, and the quantity taken was fairly small, the chances of recovery will be but slender. By far the most efficacious treatment, known at present, is that by cold affusion to the neck and spine. The following case, quoted from Taylor, will illustrate the value of this method. A girl took by mistake in medicine thirty minims of prussic acid. Immediately afterwards she became senseless; her teeth were firmly set, and her eyes fixed and staring. Stimulants failed to rouse her; the limbs were flaccid, the pupils were dilated, and she was wholly insensible. The pulse was scarcely perceptible, and the respiration was slow. A stream of cold water was allowed to fall from some height on to the region of the spine. In a minute she began to move, and became convulsed; her symptoms abated, and she became in a short time quite collected. She recovered in a few days. If the patient can swallow, the aromatic spirits of ammonia may be given, and ammonia might be applied to the nostrils; stimulating liniments to the skin have been recommended.

Tests.—There are three tests, by which prussic acid may be recognised with certainty, and they are all of great delicacy. The first depends on the formation of a cyanide of silver, by the addition of prussic acid to nitrate of silver. The dense white precipitate thus obtained may be further tested; it is insoluble in cold, but soluble in boiling nitric acid; it evolves the vapour of prussic acid when digested with hydrochloric acid; when heated in a closed tube with some crystals of iodine, iodide of cyanogen sublimes in needles, and if these are dissolved in some potash, and ferrous sulphate and hydrochloric acid added, Prussian blue is formed. The second test is the following: if prussic acid be added to a few drops of ferrous sulphate in the presence of a little potash, a dirty greenish precipitate is formed; on adding dilute hydrochloric acid to this, the solution becomes blue, and a deep blue precipitate forms, known as Prussian blue. The third test is the most delicate of all. When a mixture of prussic acid with a few drops of sulphide of ammonium is heated, the solution becomes colourless, and on evaporation leaves crystals of sulphocyanide of ammonium, the intense blood-red of sulphocyanide of iron being obtained when a neutral persalt of iron is added to this. These tests may all be performed, when the quantity of the acid is presumably small, by placing the suspected substance in a beaker, and inverting over its mouth a watch-glass with a drop or two of the test reagent in it. It is important to note, in making the examination of the contents of the stomach, whether there are any fragments of apple pips, peach kernels, bitter almonds, or any substances from which prussic acid could be obtained, which might have been taken innocently during life. At the trial of the Quaker, Tawell, for the murder of Sarah Hart, the defence was set up, that the poison had been obtained from some apple pips, a few of which had apparently been found in the stomach of the deceased; but as the quantity of the poison found was equivalent to one grain of the pure acid, this defence was entirely ignored by the jury. It is possible, however, that, in some future case, so much of the poison might have been got rid of, before the examination was made, that the hypothesis of all that was found having been derived, in some such natural way, might not be so easily refuted.

Another test, which is not applicable in the case of pure prussic acid, but will serve quite well if an alkali be added, is that a hot solution of cyanide of potassium, mixed with picric acid, gives rise to a blood-red colour, the picrocyanic acid being formed.

Turpentine.

This is liable to be taken by mistake, being an article in almost daily household use. The symptoms of an overdose may at first resemble intoxication, giving way later to narcotic phenomena. The treatment would be the same as for the other narcotic poisons, and the odour of the poison would suffice to distinguish it.

Nitro-Glycerine.

This is a dangerous explosive compound, which, when mixed with a fourth part of its bulk of earth, forms dynamite. It is liable to explode spontaneously, but is soluble in oils and fats, and is then not explosive. The symptoms of an overdose are faintness, heart failure, difficulty of breathing, and cyanosis. The blood has been found chocolate coloured in animals, which have died from the effects of it. It may be recognised by its explosive character, and by giving a red colour, when treated with brucine and sulphuric acid.

Digitalis.

All parts of the *Digitalis purpurea*, or Foxglove, are poisonous. The leaves are the most used; they are large, ovate and crenate, the young leaves being pale and silvery, the adult leaves green;

the upper surface shows a few short transparent hairs, the under surface contains many hairs. The symptoms are mainly due to the presence of an alkaloid, digitaline; but there are other alkaloids contained, viz., digitonin, digitoxin, and digitalein.

Symptoms.   When resulting from a single large dose, the symptoms come on within a very few hours. Vomiting is one of the earliest, and may be very obstinate; diarrhœa is sometimes present; the pulse is soon affected, and, though perhaps just at first it may be quickened, it very soon becomes reduced in frequency, the number of beats sometimes not exceeding twenty-five per minute. Thirst, noises in the head, suppression of urine, convulsions, dilated and insensible pupils, have all, at one time or other, been attributed to digitalis. Most frequently the symptoms are the result of repeated doses. They were well seen in two young German recruits, who, with a view to get off military service on the ground of ill-health, had been supplied with one hundred pills each by a quack to produce an abdominal affection. They began taking the pills some ten days before joining their regiment, and in two or three days one of them was taken ill, and had to be sent to the hospital, where he died after three weeks' treatment, and only thirteen pills were found remaining. His symptoms had been chiefly vomiting, with great tenderness in the stomach and a tendency to faintness, and he died from syncope in the act of going to stool. On autopsy no lesion was found to account for death, but examination of the contents of the stomach, and of the viscera, revealed the presence of powdered digitalis. The other recruit stopped taking his pills when his companion died, but he had suffered from the same symptoms, as he afterwards admitted, though to a less extent. This liability to sudden failure of the heart's action should be borne in mind in dealing with a case of poisoning by digitalis, and, indeed, in its medicinal use in large doses, and a recumbent position should be strictly enjoined. Some people are more susceptible to its effects than others. In delirium tremens enormous doses of the tincture (two ounces) have been given, without ill effects.

There are no characteristic appearances after death. Notwithstanding the vomiting, there is very often no lesion in the stomach, or only a little congestion. In the case of the recruit already quoted, the blood was dark and fluid, the right side of the heart was full, the left empty. There is no known antidote, and the treatment would consist in encouraging vomiting; solutions containing tannic acid have been recommended. *Post-mortem appearances.*

Treatment.

The chemical tests are not much to be relied upon; the best is that known as Grandeau's test; sulphuric acid and bromine give with digitaline a red colour, and with digitalein a violet colour. On the addition of water these are changed into emerald and light green. The only test, that could be really of use in the case of organic compounds, would be the physiological one of the effects produced on the heart of a frog. In a case of poisoning by digitaline which occurred in France, the evidence of the presence of an alkaloid in the vomited matters, and in the contents of the stomach, which produced the same effect on a frog's heart as a solution of digitaline did, was accepted as sufficient proof of the nature of the poison. *Tests.*

Aconite.

The root and leaves of the *Aconitum Napellus*, or Monkshood, are the parts of the plant which chiefly concern the medical jurist. The root is the most poisonous part, and its virulence is said to be increased after the leaves have fallen off. It is tapering, brown externally, internally white, and when tasted produces a cool, numb sensation at the part of the tongue where it has touched. Many fatal mistakes have occurred from the root of aconite being eaten for that of horse-radish.

Symptoms. A cool, numbing sensation in the mouth, tongue, and throat is complained of very soon; then a tingling sensation in these parts, numbness and tingling in the limbs, and loss of power in the legs. Pain in the stomach is often very acute, and vomiting and purging are common symptoms. The patient complains of dimness of vision, or actual loss of sight, and may become unconscious. In a remarkable case of poisoning in America, where liniment of aconite was swallowed in mistake, and the patient was under observation from the commencement of the symptoms, there was an extreme degree of heart failure, the pulse even in the axillary artery being quite imperceptible, though the patient ultimately recovered. The pupils are sometimes dilated. Death generally results from heart failure, and occurs tolerably early. The symptoms in all these cases are due to the existence of an alkaloid named aconitine, one of the most active poisons known. Unfortunately the different preparations of this vary very much in strength, as the following case proves. A doctor ordered for a gentleman, aged 61, some nitrate of aconitine, and French aconitine was used in making up the prescription, instead of German as had been intended. The first dose caused a feeling of constriction and burning, extending from the mouth to the stomach, and an impression of intense cold. Two hours later he took a second dose, four times as large as the first; this produced, in addition to the other symptoms above mentioned, præcordial anxiety, difficult and noisy breathing, extreme languor, cold sweats, vertigo, loss of sight, hearing and taste, and, lastly, suppression of urine, and convulsions. Spontaneous vomiting led to a remission of all these symptoms. The patient's wife went to the doctor to tell him of the dire effects of his medicine, when he, convinced of the harmlessness of his prescription, swallowed about a drachm before her, and died in five hours. At the autopsy, forty-three hours after death, rigor mortis was well marked, and there was no putrefaction. The brain and its membranes were somewhat congested, the blood in the heart and large vessels was fluid and cherry red. The stomach and small intestines, the liver, spleen, and kidneys were found to be congested.

No account of poisoning by aconite would be complete, without some reference to the murder of a boy at school a few years ago by aconitine, more especially, as it was the means of dispelling the idea that, in poisoning by the alkaloids, the medical proofs of the fact, which could be brought forward, would not be sufficiently strong to secure a conviction. Briefly, the facts of the Wimbledon murder, as it was called, are as follows: The prisoner Lamson, a medical man, called one evening at the school where the boy was, and put something, out of a capsule, into a glass of sherry that the boy was about to drink, putting in some sugar at the same time, and making some remark about counteracting the effects of the alcohol; Lamson left the room a few minutes later. Half an hour after taking the wine, the boy was seized with pain in the abdomen and vomiting, both of which continued. Some morphia was injected over the epigastrium, but the boy's sufferings steadily increased, and he died two hours and a half after taking the wine. At the post-mortem examination

The trial of Lamson.

on the next day, no lesions of any moment were found, except some recent inflammation of the stomach; no natural cause for so rapid a death, from a state of previous good health, was discerned, and all that could be said was, that the boy had died from the effects of an irritant poison, which was probably of a vegetable nature. The vomited matters, the urine, and portions of the viscera, were preserved for analysis. Dr. Stevenson, who had performed the analysis with Dr. Dupré, described the steps he had taken, when in the witness-box. In the first place, a solid extract was made from each tissue or substance. As these extracts did not give any of the chemical reactions of the known alkaloids, Dr. Stevenson tasted each of them, and described the effect as a peculiar sensation, a burning, tingling, a kind of numbness difficult to define, a salivation creating a desire to expectorate, a sensation at the back of the throat of swelling up, followed by a peculiar seared sensation of the tongue as if a hot iron had been passed over it, or some strong caustic placed upon it. In some instances the sensation was much less marked, there being merely a faint tingling; the sensation on the tongue lasted for varying lengths of time, the longest being seven hours. Some of the urine was injected beneath the skin of a mouse, which was obviously affected in two minutes, and died in half an hour. Some of Morson's aconitine produced exactly the same symptoms, when injected into a mouse. When the extract made from the vomit was injected the mouse died in a quarter of an hour. The presence of aconitine in the tissues of the body, and in the urine and vomit, having been thus conclusively demonstrated, the prisoner was convicted. The defence was exceedingly weak. An attempt was made to show that the symptoms, with which the boy died, were attributable to an old standing disease of the spine, for which there was not the slightest foundation; and next it was argued that the alkaloid, which Dr. Stevenson had found, was really a cadaveric alkaloid, or ptomaine. This theory Dr. Stevenson was able to dispose of in a very few minutes; in the first place, the boy had not been dead long enough for the development of any cadaveric alkaloids, and, in the second place, the reduction of the ferricyanide of potassium to ferrocyanide, which always takes place in the presence of a cadaveric alkaloid, was not obtained. Evidence was given of the purchase of aconitine by Lamson, at a leading chemist's, and as the boy was his brother-in-law, and was possessed of some money—a portion of which would come to Lamson's wife in the event of the boy's death—no link in the chain of moral, or medical, evidence was wanting.

The recital of these cases has made it pretty clear that there are no post-mortem appearances, in cases of this kind, which have any diagnostic value; neither are there any chemical reactions by which aconite can, at present, be recognised. In the present state of knowledge, experiments on animals for the detection of the poison, when aconite has been used, can alone afford reliable evidence. The treatment recommended is to give emetics, to wash out the stomach, and to keep the patient in the recumbent posture. Hypodermic injections of atropine or digitaline might be used, as also the inhalation of nitrite of amyl.

CHAPTER XXVIII.

POISONOUS GASES.

The chief gases that come under this head are carbonic acid, carbonic oxide, coal gas, and sulphuretted hydrogen.

Carbonic acid, or carbon dioxide, is present everywhere in some degree, but is not perceived, unless the proportion of it becomes excessive. In a crowded and ill-ventilated room, it soon accumulates to the point of producing unpleasant symptoms. It collects in wells, mines, and cellars, and is given out in large quantities during the burning of lime in kilns, and being heavy it tends to sink towards the ground; owing to this fact many lives have been lost of persons incautiously sleeping too near lime-kilns. If present in a large quantity it renders the air quite irrespirable, and the person placed in such an atmosphere would rapidly be suffocated. When not so concentrated, it produces headache, drowsiness, giddiness, loss of power, the pulse and breathing are hurried, and the body-surface becomes livid, coma supervenes, and the person dies. The usual post-mortem appearances are congestion of the brain and its membranes, in addition to the ordinary signs of asphyxia, *i.e.* lividity of the face, protrusion of the tongue, frothy mucus about the mouth, and great engorgement of the right heart and venous system.

The blood is dark and fluid. The treatment would be to employ artificial respiration, and to let the patient have plenty of fresh air.

Carbonic oxide acts as a poison to respiration, and, when inhaled, soon leads to unconsciousness and death. A few years ago two cases of very sudden death from poisoning by this gas occurred, at a paper manufacturer's in Germany. A workman having put some lime into a huge kettle, before the rags to be acted upon were placed there, was poisoned by the gas thus given out, when he got into the kettle to tread the rags down in the usual manner. A fellow workman, who, when this was found out, got into the kettle and handed out the body, was also poisoned in the same way and died. The explanation of the development of the gas was, that owing to the presence of moisture in the kettle, when the unslaked lime was introduced, either contained in the rags or left from the previous cleaning, hydrate of calcium had been formed with the development of a great amount of heat. The rags would thus have been subjected to a sort of incomplete combustion, and such gases as carbonic monoxide and marsh gas would have been formed. On both the bodies there were characteristic light-red blotches. The blood and the muscles were a light red colour, notwithstanding that decomposition was far advanced.

It is supposed to act by combining with the hæmoglobin of the blood and forming a tolerably fixed compound. On examination of the blood the absorption band of the hæmoglobin is found nearer to the red end of the spectrum than it is under normal conditions. When treated with caustic soda the blood gives a red precipitate, passing into a red solution. Transfusion of arterial blood is said to be the only means of saving the patient's life.

Coal gas, or illuminating gas, is a compound of

a great many gases. Headaches, giddiness, loss of consciousness, convulsions, are the effects of the admixture of this gas with the ordinary air in a room. Many times gas has escaped into a bedroom during the night, and caused death without the person being aroused by its effects. Pettenkofer has recently called attention to the state of illhealth, which is liable to be produced by the gradual leakage of gas through the floor of a room, or house, without the cause being suspected. He suggests further that the symptoms may be somewhat masked, owing to modifications which the gas may have undergone in its passage through the earth. The symptoms are said to resemble those of apoplexy, and to differ from them chiefly in being more fluctuating. The appearances after death are injection of the papillæ at the base of the tongue, and froth in the air passages; the lungs are of a brilliant colour, and sometimes extravasations of blood in the spinal cord or membranes are found.

Sulphuretted hydrogen produces in the first instance nausea, drowsiness, giddiness, pain in the stomach, then irregular action of the heart, later on coma and death, with tetanic symptoms or convulsions; when very concentrated, death may ensue very rapidly. The mildest degree of poisoning would be a febrile state with symptoms referable to disorder of the intestines. Putrefaction takes place rapidly in the dead body; the blood is fluid and dark coloured. The larger bronchi appear as if smeared with a brownish deposit, which can easily be wiped off. This is considered very characteristic. The treatment would be to get the patient in a current of fresh air, and apply cold affusions, giving stimulants internally.

CHAPTER XXIX.

PREGNANCY.

THE symptoms and signs of pregnancy are numerous; taken individually, there is hardly one on which absolute reliance could be placed, supposing the others to be absent; but when the majority of the more important are present, doubt is no longer possible. Taking them in order of their appearance, and dealing first with those that concern the mother, they may be arranged in the following manner, as given by Dr. Tidy: 1. Cessation of the catamenia. 2. Morning sickness. 3. Fulness of the breasts. 4. Enlargement of the abdomen. 5. Changes in the uterus. Those more especially having reference to the fœtus are two in number: 1. Quickening. 2. The sounds of the fœtal heart.

In the vast majority of women, after conception has taken place, the menses cease, and do not reappear during the whole time of pregnancy. But this is not universal, and their return on one or more occasions at the usual time, after conception, may throw a woman very much out in her calculations as to the stage of her pregnancy. Menstruation may indeed continue throughout the whole of the pregnancy, and the fact of its persistence must not be used as an absolute proof of the non-existence of pregnancy in a doubtful case. On the other hand, women have been known to go on bearing

Cessation of the catamenia.

children, without the reappearance of the menses in the intervals of each pregnancy; this is not, therefore, a necessary condition for impregnation. As a rule, the health of a girl suffers obviously when the menses are arrested from disease, so that the continuance of good health, with their non-appearance, would often suggest the possibility of pregnancy. It must not be forgotten, too, that menstruation may be feigned, in order to conceal the fact of pregnancy.

Morning sickness. In the course of the second month morning sickness usually comes on, and is rarely absent in primiparæ; it may occur as early as the second week after conception; it usually lasts till about the fourth month, disappearing about the time of quickening. As the name implies, it is most constant in, and indeed is often limited to, the morning. As a rule, the general health does not suffer, and the appetite remains good; occasionally it is so obstinate as to be a symptom causing some alarm, but it hardly ever, if at all, leads to abortion. The symptom might be due to chronic intemperance or hysteria, but the supervention of other symptoms would soon aid in clearing up the diagnosis in a doubtful case.

Changes in the breasts. Between the second and third months of pregnancy the breasts first begin to be prominent and full. The areolæ begin to show changes after the second month: they become larger and darker, the skin is always soft; the glandular follicles in the areola are somewhat prominent, and secrete a little fluid. These signs are of more value in a primipara, for in a woman who has already borne children the areolæ do not return to their original condition. In the later stages of pregnancy, on squeezing the nipples and breasts, a little transparent fluid can be obtained. It must be noted, however, that in a woman, who has already borne children, the secretion of milk may be reproduced under various and often trivial circumstances, whilst enlargement of the breasts may follow on certain ovarian or uterine disorders. A discolouration of the areola may arise, too, apart from pregnancy, so that, although this condition of the breasts is a sign of some value, too much stress must not be laid upon it.

Towards the middle or end of the third month the contents of the uterus attain such a size that the uterus rises out of the pelvis, and causes a visible prominence of the abdomen, the enlargement of which, from that time forward, steadily increases. The degree of enlargement, and the extent to which it is noticeable, vary in different individuals, and may be modified by the efforts made to conceal the fact of pregnancy; still, enlargement of the abdomen is a sign which can never be wholly absent. Its value as a sign of pregnancy is, of course, greatly diminished by the fact that there are many diseases associated with enlargement of the abdomen, so that, by itself, it ought hardly to arouse even a suspicion of pregnancy. The points of diagnosis between the various forms of abdominal tumour, ascites, and ovarian dropsy will be found in the text-books on medicine, and need not be described here; but the possibility of the coexistence of pregnancy with ovarian disease should be borne in mind. In all such cases a good deal must depend upon the statements of the woman herself, as to whether the pregnancy will be overlooked or not. A brown streak down the abdomen, from the umbilicus to the pubes, has been described as indicative of pregnancy; but as it is present in enlargement of the abdomen from other causes, no stress should be laid upon its existence. The umbilicus gradually becomes less and less depressed as the pregnancy proceeds, becoming quite flattened out about the eighth month, and after that forming a prominence. If the hand be laid upon the abdomen, after the uterus has risen out of the pelvis, a distinct contraction may be felt, especially if the precaution of dipping the hand into cold water has been taken. A tumour, which can be felt to contract under the hand in this way, can be nothing else than the uterus, though the sign is not in itself a proof of pregnancy, but only of an enlarged uterus.

<small>Changes in the uterus.</small>
During the first few weeks after conception the cervix will be found projecting into the vagina, of its usual length, but perhaps somewhat tumid. After the uterus has risen from the pelvis, according to some authorities, the cervix becomes shorter; according to others, it does not show much change until the last few weeks of gestation, when it gradually becomes spread out like a thin membrane. When the abdomen is palpated, as mentioned in the last section, during the intervals between the contractions of the uterus, the body of the fœtus may be plainly perceptible, and sometimes its movements can be recognised. After the sixth month, another sign may be recognised, known as *ballottement*. It consists in giving a sudden impulse through the cervix to the fœtus with the finger of one hand in the vagina, the other hand remaining flat upon the abdomen, when the sensation will be felt of a foreign body coming up against the hand, and if the manœuvre be performed with the person in the upright position, the return of the fœtus against the finger will be felt. This sign, with the others that have preceded, is conclusive evidence of pregnancy, but it does not prove that the fœtus is living.

One other sign may be mentioned, though it has no value. It is the presence of what has been called 'kiestein' in the urine, which consists of an amorphous pellicle on the surface of the urine, appearing about twenty-four hours after it has been passed. It is no longer regarded as any indication of pregnancy, and no importance should be attached to its recognition.

As already mentioned, there are two signs which belong more properly to the fœtus. The first of these, that known as *quickening*, takes place somewhere between the twelfth and sixteenth weeks. It consists in certain peculiar sensations felt by the woman, of movements within her, and is often associated with nausea, faintness, and other unpleasant symptoms. If sickness has been present, it usually ceases when quickening is past. The meaning of the word 'quickening' is living, it being formerly supposed that when these sensations occurred the fœtus received vitality for the first time. It is on this hypothesis that so much stress is laid on it by the law; so that, legally, the mere fact of the woman being pregnant is not inquired for by the law, but only whether she has quickened or not. Quickening is now regarded as due to the fœtal movements, at the time when the uterus rises out of the pelvis, and is in part, perhaps, also due to the change in position of the uterus, which cannot fail to exert a marked influence on the sympathetic nervous system. The movements of the fœtus will not be perceptible by examination, in the manner already alluded to, until after the period of quickening, so that their recognition is in itself a proof that the time of quickening is past, but the fact that they cannot be so felt is no argument against the quickening having taken place. Nothing but the woman's own statements can afford satisfactory evidence under such circumstances. Very few women can tell the exact date on which they first felt the sensations, so that the occurrence of quickening cannot be relied upon as an indication of the date of probable delivery. Some women never notice the sensations at all, others mistake other sensations for them.

Quickening.

310   MEDICAL JURISPRUDENCE.

Sounds of the fœtal heart.

The second sign consists in hearing the *sounds* due to the beating of the *fœtal heart*. This sign would in itself be conclusive, not merely of the existence of pregnancy, but that the fœtus was living; but, of course, it can never exist without the other signs, viz. enlargement of the abdomen, and, probably, the changes in the uterus and the movements of the fœtus, being readily demonstrable. The point at which the sounds are most likely to be heard, when the fœtus is in the position it most frequently occupies, is about midway between the umbilicus and the anterior superior spine of the left ilium. If they cannot be heard at this spot, they may be heard, assuming the fœtus to be in the next most common attitude, at the corresponding spot on the right side of the abdomen. If the fœtus should not happen to be in either of the most common positions, it is possible that the sounds will not be audible anywhere. There are other reasons, too, which might combine to prevent their being heard, so that this fact is no proof that the fœtus is not alive. If the sounds are found to be synchronous with the mother's pulse, they are not due to the fœtal pulsations. Their frequency is said to bear an inverse ratio to the stage of pregnancy, being about 160 at the fifth month and about 120 at the full term. It has also been said that they are more frequent in the case of a female than a male fœtus. Besides these sounds, there is another murmur, or bruit, often audible, which is supposed to be produced in the uterine sinuses, and known as the placental bruit or uterine souffle. It is, however, also present in cases of uterine fibroids, and therefore has no diagnostic value.

Pregnancy in the dead.

If the pregnancy was so far advanced that ossification had commenced, then the bones of the fœtus might be discovered at almost any period after death. In the recently dead body, the discovery of a fœtus, or embryo, would of course be conclusive. The value of the presence of a corpus luteum will be discussed subsequently. In dealing with the subject of the identity of the dead, it was mentioned that the virgin uterus has a great power in resisting decomposition. If, therefore, in the body of a woman who has been dead for a year, the uterus is found to be tolerably firm and has maintained its colour, it will be certain that it is the uterus of a woman who has not borne a child.

A woman who has not borne children will have greater difficulty in feigning pregnancy, than one who has had the advantage of that personal experience. In either case, if the advanced stage is professed, when the woman is examined, it is not possible that the fraud should be undetected, unless she should happen to have some form of ovarian disease. In such a case, a medical man might feel that there was a possibility that the signs of pregnancy were so masked by those of the ovarian disease, that he would not commit himself to a positive opinion. If the woman refused to submit to an examination, the natural inference would be that her story was untrue. A medical man must never forget that he may not examine a woman without her consent, except by the order of a judge after a criminal trial. To do so would be to render himself liable to an action for assault. Putting aside instances where pregnancy is alleged to support a charge of rape, or to please a husband, there are two cases in which a question as to the reality of an alleged pregnancy may have to be determined. One of these is of a civil nature: A woman, at the death of her husband, may declare that she is pregnant with an heir to his estate, she having been hitherto childless, and the heir-at-law may sue out a writ to require proof of her alleged pregnancy. The other case, where the reality of a

Feigned pregnancy.

pregnancy has to be inquired into, relates to the criminal law. If a woman is convicted of a capital offence, and sentenced to death, she may plead pregnancy as a bar to execution. A jury of twelve women is then empanelled from those in court (for the plea must be raised at once), to try whether the prisoner be 'with child, of a quick child or not,' to use the actual words of the statute. The women from whom the jury is of necessity selected are, as a rule, ignorant persons, and, as a matter of fact, one or more medical men are always ordered to assist them in coming to a conclusion; for the great probability is, that none of the jury would be competent to do more than form an opinion, by handling the abdomen externally—a most unreliable method by itself. Moreover it is a proof of quickening that is required, and not of pregnancy; and it has already been shown that, in many cases, this cannot be proved, but can only be inferred from the statements of the woman herself, who must always be a prejudiced witness in such a matter. In America, questions of this nature are referred to a committee of six physicians, who have power to subpœna witnesses.

Concealment of pregnancy.

In Scotland it is an offence against the law to conceal the fact of pregnancy, and the woman who does so is liable to be prosecuted, if her child dies. The idea is, that every woman, who is pregnant, is bound to make preparations for the safe delivery of her child, and if the child is born dead, and in secret, it is assumed that its death is the result of the want of such preparation. Evidence of the actual delivery of a child is required before conviction.

Unconscious impregnation and unconscious pregnancy.

The possibility of impregnation taking place whilst a person is under the influence of an anæsthetic, a narcotic, or alcohol, cannot be denied, though it is difficult to believe that a virgin on awaking should not even suspect that anything had occurred. But all cases of alleged impregnation, when the female was in a state of unconsciousness, should be received with a great degree of caution; the majority are probably untrue. In a very few cases married women, who have already borne children, have become pregnant and gone to the full term without being aware of their condition; but this must be regarded as quite exceptional.

CHAPTER XXX.

DELIVERY.

In cases where there is any reason for concealing the fact of delivery, the same reasons will have necessitated the concealment of the pregnancy, and as the woman will not be likely to say anything which would tend to criminate herself, the only means of judging as to whether delivery has, or has not, taken place, will be by an examination of the woman during life, or of her body after death.

Recent delivery in the living.
In medical language, if the contents of the uterus are expelled before the end of the sixth month, the woman is said to abort or miscarry; if after the sixth month, but before the completion of the full term, she is said to have had a premature confinement, and the child may or may not be still-born. The law, however, takes no notice of these artificial distinctions, but classes them all under the head of delivery. It is obvious, therefore, that the signs of delivery will vary according to the state of gestation, that had been reached before the birth took place. If this happens before the third month, it is clear that many of the symptoms and signs, which have been described as evidences of pregnancy, cannot be present, as they would not have had time to make their appearance. Such are, amongst others, the changes in the breasts and enlargement of the abdomen. At this early stage,

then, proof of abortion is very difficult; the woman would probably lose a considerable quantity of blood, and the discovery of an ovum, or any of the membranes, would be the only clue to the fact that an abortion had taken place. The embryo at this stage is so small that no signs of its passage will be discovered on examining the os uteri, though at a somewhat later period some local discharge may be set up. Unless the examination is made within twenty-four or thirty-six hours of the occurrence, there would be extremely little likelihood of any traces of the abortion being recognised.

Ogston thus sums up the signs of recent delivery in the living: 1. The proof of recent parturition in the female can only be safely based on the presence of most, if not all, the signs enumerated. 2. These signs are found to be more or less marked and durable according to the greater or less severity of the previous labour, and the greater or less vigour of the woman's constitution. 3. In general, these signs cease to be distinguishable after the eighth or tenth day. 4. The fact of actual delivery may be considered as demonstrated, and the time at which it had taken place, fixed at two or three days preceding the examination, if the breasts are found enlarged, if the superficial mammary veins are prominent, if milk of a yellowish colour resembling serum, and of a disagreeable taste (colostrum), can be pressed out of the nipples; if the abdominal parietes are relaxed and covered with minute pink cracks; if the recti abdominales are widely separated; if the vulva be gaping, contused, and swollen; if the fourchette be recently ruptured; if the vagina be wide, with its rugæ obliterated; if the os uteri be open, and its lips pendant and thickened; if the uterus form a bulky tumour in the hypogastrium; and finally, if a sero-sanguinolent discharge be issuing from the vulva. 5. The

Ogston's signs of recent delivery.

delivery may be assumed to have taken place three or four days previous to the examination, if the traces of contusion and distension of the external genitals, though still distinguishable, are less evident; if the sero-sanguinolent discharge have ceased, or is very slight; if there be feverish symptoms; if the sweat have the [characteristic] odour; and if a milky serum be found in the breasts, indicating that the milk fever is at its height, or beginning to decline. 6. The delivery dates from at least five or six days to at most eight or ten days, when the contusion and distension are no longer distinctly perceptible; when the uterus has shrunk in the hypogastrium, and can still be felt as a small rounded tumour; and when the lochial discharges are thick, yellow, and very fœtid. 7. If there be no trace of contusion or distension; if the lochia are watery, and nearly inodorous, and if the uterus is scarcely to be felt, it may be assumed, with some probability, that the period of the delivery has preceded that of the examination by about fifteen days.

Signs of delivery at a remote period.

When a question is raised as to the recognition of the fact of previous delivery at a remote period, there can be little doubt that, unless the delivery took place at or near the full term, the chances are that the signs remaining could not be recognised. In the case of delivery at the full period, the most valuable signs would be scars about the perinæum, vagina, or uterus; the absence of these is, of course, no proof that delivery has not taken place at some anterior period. Other signs, upon which a varying degree of stress has been laid at different times, are, silvery lines on the breasts, the lineæ albicantes on the abdomen, the enlarged areola round the nipple, and the brown mark from the pubes to the umbilicus. Although it can be said of any, or all, of these, that they may result from other causes than previous delivery, yet a satisfactory explanation should always be demanded, before deciding that they were not the effects of delivery. Moreover, their absence is not a proof that delivery has not taken place, so that it is evident that the question as to whether a living woman has or has not, at some prior time, been delivered of a child, at or about the full term, does not, in every case, admit of a decisive answer. The condition of the os uteri has been thought by some to afford valuable evidence as to the fact of a previous delivery, but it varies very much in those who have not borne children, and even at each menstrual period. Dr. Tidy proposes rupture of the posterior commissure as a reliable guide of delivery having taken place, and its non-rupture as equally valuable evidence that delivery has not taken place.

Feigned delivery can only be investigated and exposed when recent; if the woman submits to an examination the fraud must be detected; if she refuses to be examined, her statements will be regarded with suspicion. *Feigned delivery.*

In cases of infanticide, it occasionally happens that the woman states that she was quite unaware of her delivery, at the time it occurred. Delivery may, no doubt, take place unconsciously, under the influence of an anæsthetic, or of a narcotic, or of alcohol; but, except under such conditions, it would seem impossible that a woman, especially if a primipara, should pass through the pangs of labour unconscious of them. In women who have borne several children, who have an unusually wide pelvis, and especially when the child is small, parturition may take place so rapidly, and with such ease, as almost to be free from pain. Under such circumstances it is possible that a child might be born whilst the mother was straining at stool, as is *Unconscious delivery.*

not infrequently alleged in cases of child-murder, but that the mother should not instantly after be aware of what had happened is not to be believed. Delivery might take place without the mother being conscious of it under two other circumstances, viz. if she was insane or was suffering from epilepsy or eclampsia; the discovery of albumen in the urine would go far to support a woman's statement, that she had been unconscious at the time of her delivery.

Post-mortem delivery.

There are several well-authenticated instances of parturition after death. Formerly such an event was regarded with much awe and superstitious horror, but it is now generally considered to be the result, partly of a natural contraction of the uterus, and partly of the accumulation in the abdominal cavity of gases due to putrefactive changes.

Recent delivery in the dead.

The signs of recent delivery in the dead have next to be considered. These will depend, of course, upon the stage at which gestation had reached, and will be slight and evanescent in proportion to the immaturity of the fœtus. Thus, after a six months' miscarriage, if death had not taken place for three or four days, all traces would probably have disappeared. If the delivery occurred at the full term, and the death not till about the tenth day, the signs will, in all likelihood, not be sufficiently decisive to justify an opinion as to the probable date of delivery. As the cases, in which it is of importance to determine the date of delivery, are mostly of a criminal nature—*i.e.* cases of abortion—the fact of the pregnancy will probably be concealed, and no aid be derivable from the history of the case. When the death has taken place soon after the delivery, the uterus will be found flattened like a great pouch, with a gaping mouth, and measuring from nine to twelve inches in length. Clots of blood, or a blood-stained fluid, will be found in its interior, and likewise the remains of the decidua adhering to the walls, except at the point where the placenta was attached, which is dark coloured, and looks raw, showing the semilunar openings of the uterine sinuses. At this spot the discolouration is so marked as to suggest that decomposition has begun, or gangrene set in. The Fallopian tubes, ligaments, and ovaries appear very vascular, and the spot whence the ovum escaped is more purple than the rest of the organ.

As regards the size of the uterus at the various periods after delivery, there has been much difference of opinion. Dr. Montgomery, one of the leading writers on obstetrics, says: 'But when delivery has taken place at full time, and the uterus has contracted perfectly, if an examination be made within a day or two, it will be found about seven or eight inches long and four broad, its external surface having a vascular appearance, and not unfrequently presenting patches of a purplish colour; its substance, divided by the knife, is found from one to one and a half inches thick, of the consistency, and nearly of the colour, of firm muscular fibre, of which it appears to consist, and the cut surface displays the orifices of a great number of very large vessels; it now weighs about a pound and a half. . . . At the end of a week the organ has diminished to between five and six inches in length, and weighs about one pound and a quarter; after a fortnight it does not exceed five inches in length, and its weight is reduced to about three-quarters of a pound, or a little less; its vascularity is diminished, and the thickness of its parietes is reduced about one-third; their density is, however, increased in like proportion, so that the orifices of the vessels are much less distinct, and the colour of the muscular substance has become much paler.' The uterus will not, as a rule, be fully contracted in less than a

[margin: Evidence from the state of the uterus.]

month from the date of delivery. As to whether, or not, it completely returns to the state it was in before conception, will be shortly discussed. As already stated, the appearances will depend upon the stage of gestation, and in an early stage they may be very slight. It would be well to bear in mind the condition found in the bodies of women who have died during menstruation : the uterus is a little enlarged, its walls being thickened and its mucous membrane lined with a vascular layer; beneath this the substance of the uterus is white and firm. After delivery, there is generally some bruising found about the neck of the uterus.

Corpus luteum.

A great deal of importance has been attached to the discovery of a corpus luteum in the ovary, in former times, as an evidence of impregnation, but this is not now considered to have the same significance that was formerly attached to it. A corpus luteum is the cicatrix on the ovary at the spot whence the ovum has escaped. It consists of a swelling, oval in form. and of a yellow colour. On section it is found to be full of blood, and to have a central cavity, or a stellate cicatrix, according to the date at which the examination is made. The corpus luteum gradually disappears, so that by the time gestation is completed it has nearly vanished. The scar left by the unimpregnated ovum sometimes closely resembles it, so that its presence cannot be regarded as an infallible proof of conception having taken place ; and further, the ovum may become impregnated after it has left the ovary, so that the absence of a corpus luteum is not a proof that conception has not occurred. The corpus luteum has therefore been spoken of as true or false, according as it is the result of conception, or simply of menstruation.

Tidy contrasts their characters in a table as follows :

DELIVERY.

At the end of—	True	False
Three weeks.	Three-quarters of an inch in diameter. Central clot of a reddish colour. Convoluted wall pale.	
One month.	Larger size. Convoluted wall bright yellow. Central clot continues of a reddish colour.	Smaller. Convoluted wall of a bright yellow. Clot continues of a reddish colour.
Two months.	Seven-eighths of an inch in diameter. Convoluted wall bright yellow. Central clot perfectly decolourised.	Reduced to the condition of an insignificant cicatrix.
Six months.	As large as at the end of the second month. Convoluted wall paler, but still of a yellow colour. Central clot appears fibrinous.	Entirely disappeared.
Nine months.	Half an inch in diameter. The external wall tolerably thick, and convoluted, but the yellow colour disappeared. Central clot converted into a radiating cicatrix.	

It is evident that, whilst so much uncertainty prevails on this subject, the presence of a corpus luteum cannot be utilised in evidence in a court of law.

The signs, discussed up to the present, have related to the determination of recent delivery in the dead body, but it may be necessary for purposes of identity, or for other reasons, to determine whether delivery has taken place at any previous time. If the body be only recently dead, the condition of the breasts should be noticed, for although a large

Evidence of delivery at any previous date.

322 MEDICAL JURISPRUDENCE.

and pendulous state does not prove pregnancy at some antecedent date, yet the discovery, that the breasts were in their virgin condition, would afford strong presumptive evidence, that the body in question was that of a woman who had never borne a child; the same reasoning will apply to the lineæ albicantes. But the external parts of the body may be so altered by decomposition that these points cannot be made out, and the whole question will then turn upon the state of the uterus. It has already been shown that the virgin uterus resists decomposition better than the uterus which has been pregnant. When a long time has elapsed between the pregnancy and the death, the difficulty of deciding may be very great, but the uterus does not quite return to its original size after delivery; the cavity is a little larger, and the walls remain thicker than before. The neck of the uterus, in those who have borne children, is shorter than it is in the virgin state. But these differences are only relative, and they are therefore the less useful as being likely to be appreciated at different values by different observers. In the Wainwright trial the majority of medical witnesses were strongly of opinion that the body of the murdered woman was that of one who had borne children, but there were some witnesses who held that an absolute opinion on such a point could not be given. Amongst other signs, which have been mentioned by various writers, may be noted a thinning of the centre of the flat portions of the iliac bones, and a general tendency to loosening of the joints of the pelvis.

Characters of the embryo.

The discovery of an ovum or embryo, or fœtus, in the cavity of the uterus affords the best proof of pregnancy. For the characters by which the ovum may be recognised in the first two months after conception, the student is referred to the text-books on physiology; suffice it to say that at the sixth week it is about six lines long, and at the end of the second month about three-quarters of an inch in length. At this time the ovum consists of the embryo and its membranous coverings; the presence of the decidua and chorion may be recognised, even if the embryo should not be found; the placenta begins to assume its regular form about this time. From this date development proceeds rapidly. At the third month the embryo is from two to two and a half inches long, and weighs from 480 to 720 grains. It is henceforth called the fœtus, and at the fourth month its length is five to six inches, and its weight two to three ounces. At the fifth month its length is six to seven inches, and its weight five to seven ounces. At the sixth month it measures from nine to ten inches long, and weighs one pound. After this date the fœtus becomes viable, and the characters it presents at the several stages between that date and the full time will be subsequently discussed.

The concealment of the birth of a dead child, or of one which dies before its birth has been made known, is an offence against the law. The statute relating to this subject runs as follows:—If any woman shall be delivered of a child, every person, who shall by the secret disposition of the dead body of the said child, whether such child died before, at, or after its birth, endeavour to conceal the birth thereof, shall be guilty of a misdemeanour, and being convicted thereof, shall be liable at the discretion of the Court to be imprisoned for any term not exceeding two years, with or without hard labour. A proviso is added to the effect that any person, tried for the murder of a child and acquitted, may be found guilty of concealment of birth, if it shall appear in evidence that the child had been recently born, and that such person did, by some secret disposition of the dead body, endeavour to conceal the birth.

<small>Concealment of birth.</small>

CHAPTER XXXI.

ABORTION.

Definition. THIS is the name given by the law to the expulsion of the contents of the uterus at any time before the full term of gestation has been reached. In medical language abortion is only used in reference to the expulsion of the fœtus before the sixth month of gestation, subsequent cases coming under the head of premature labour. Criminal abortion is not often attempted before the third month, as up to that time the woman will have had no certainty of her condition. From that date to the end of the fifth month is the most common time for abortion, as the woman feels no longer any doubt as to her pregnancy, and is especially anxious to be rid of the fœtus before her condition is discovered by those about her.

Statutes relating to criminal abortion. The following are the statutes (24 & 25 Vict. cap. 100, sects. 58, 59) relating to the subject of criminal abortion :—1. Every woman being with child who, with intent to procure her own miscarriage, shall unlawfully administer to herself any poison or other noxious thing, or shall unlawfully use any instrument or other means whatsoever with like intent, and whosoever, with intent to procure the miscarriage of any woman, whether she be or be not with child, shall unlawfully administer, &c., shall be guilty of felony. 2. Whosoever shall unlawfully supply or procure any poison or other noxious thing, or any instrument or thing whatsoever, knowing that the same is intended to be unlawfully used or employed with intent to procure the miscarriage of any woman, whether she be or be not with child, shall be guilty of misdemeanour, and being convicted thereof, shall be liable, at the discretion of the Court, to be kept in penal servitude for the term of three years, or to be imprisoned for any term not exceeding two years.

Criminal abortion not unfrequently terminates in the death of the woman, and in such a case the person who procured the abortion will be liable to be put on his trial for murder. At the trial a few years ago of a German doctor, an ingenious attempt was made to reduce the crime to manslaughter, on the ground that the instrument used was not in itself a dangerous one, and that there had been no intention to destroy life. The judge who tried the case expounded the law as follows: ' If a man for an unlawful purpose used a dangerous instrument, or medicine or other means, and thereby death ensued, that was murder, although he might not have intended to cause death, although the person dead might have consented to the act which terminated in death, and although possibly he might very much regret the termination which had taken place, contrary to his hopes and expectations. This was wilful murder. The learned counsel for the defence had thrown on the judge the task of saying whether the case could be reduced to manslaughter. There was such a possibility, but to adopt it would, he thought, be to run counter to the evidence given. If the jury should be of opinion that the prisoner used the instrument not with any intention to destroy life, and that the instrument was not a dangerous one, though he used it for an unlawful purpose, that would reduce the crime to manslaughter. He really did not think they could come to any other conclusion than that the instrument used was a dangerous one, if used at all. Then if it were so used by the prisoner the case was one of murder, and there was nothing but a verdict, either of murder or acquittal.'

From the statutes above quoted it is evident that it is not necessary that the woman should be pregnant; the intent of the person in administering the drug, or using the instrument, is all that the law looks to, and as it is not necessary that the woman should be pregnant, it follows as a matter of course that it is not necessary to the crime that abortion should have been procured, or even that the means adopted should have been able to procure abortion; so long as the person using those means believed that they would have such an effect, it matters not what their real effect would have been. The use of the phrase 'noxious thing' has often been the cause of discussion in the law courts. In a very recent case the Court decided that 'in each case it was a question for the jury to say whether the substance, administered as it was, and under the circumstances in which it was administered, was a noxious thing. Therefore neither principle nor authority preclude us from holding what is certainly good sense, that if a person administer, with intent to produce miscarriage, something which as administered is noxious, he administers a noxious thing.'

It is obvious, however, from the way in which the statute is worded that it will not be necessary, in future cases, to prove that the substance administered was a noxious thing, for the words 'or any other means whatsoever' are most comprehensive. This view was taken some years ago by a judge who, in addressing a grand jury in reference to a case of abortion, called their attention to this particular paragraph, and said that he should hold that if the person accused did administer something, which he thought would procure miscarriage, with that intent, although the thing itself would not procure that miscarriage, he would be guilty of the offence.

It is unfortunately only too true that medical men have allowed themselves, on rare occasions, to be enticed into procuring abortion, and such cases, when brought to light, are justly punished with unusual severity. The mere giving a prescription to a woman who alleged herself to be pregnant, that she might try to procure abortion for herself, was, in one instance, followed by a sentence of five years' penal servitude, though the woman was not pregnant, was notoriously of bad character, and had only obtained the prescription with the view of extorting money.

The causes of abortion may be divided into the natural and the violent. Of the former syphilis is the most important, and it is a common thing for a woman whose husband has had syphilis, to miscarry before she has any children. Mental emotion may bring on an abortion, and when a woman has once miscarried there is a certain tendency to miscarry again, at the same period of her next pregnancy. Acute diseases, such as variola or typhoid fever, are almost certain to produce miscarriage. There are certain diseases of the placenta, too, apart from syphilis, which induce miscarriage.

Causes: Natural.

Dr. Robert Barnes divides the natural causes of abortion into those dependent on the mother, and those dependent on the fœtus. The *maternal* causes are:—1. Poisons circulating in the mother's blood: (*a*) Introduced from without, as fevers, syphilis, various gases, lead, copper, &c. (*b*) Products of morbid action, as jaundice, albuminuria, carbonic acid from asphyxia, and in the moribund. 2. Diseases degrading the mother's blood: anæmia, obstinate vomiting, over-lactation. 3. Diseases disturbing the circulation dynamically (mechanically), as liver, heart and lung disease. 4. Causes acting through the nervous system : (*a*) Certain nervous diseases, as chorea, &c. (*b*) Mental shock. (*c*) Diversion or exhaustion of nerve force, as from obstinate vomiting. 5. Local diseases: (*a*) Uterine diseases, as fibroid tumours, inflammation, hypertrophy, &c., of the uterine mucous membrane. (*b*) Mechanical anomalies, as retroversion, pressure of tumours external to uterus, &c. 6. Climacteric abortion. 7. Abortion artificially induced. The *fœtal* causes are:—1. Diseases of the membranes of the ovum : fatty degeneration, hydatidiform degeneration, inflammation, congestion, apoplexy, and fibrous deposits. 2. Diseases of the embryo itself : malformation, inflammation of the serous membrane, diseases of nervous system, diseases of kidneys, liver, &c., and mechanical, as from twisting of the cord or funis. The cause may be partly maternal and partly fœtal.

Violent.

The violent causes are of two sorts, viz.: 1. Mechanical. 2. Medicinal. Of the former, severe exercise and rude shocks to the body, especially to the abdomen, are often resorted to; people have thrown themselves out of window to effect their object; repeated bleedings might also be mentioned in this category. Blows on the abdomen, and even trampling on the abdomen, have been known to fail in attaining the wished for result. The most effectual means for procuring abortion, and one well known to all abortionists, is the rupturing the membranes and allowing the escape of the waters. When this has taken place the death of the fœtus is assured, and its expulsion, at no very distant date, follows as a matter of course. The induction of artificial labour by this method, however, is an operation requiring a knowledge of the anatomical relations of the parts involved, which as a rule the abortionist does not possess, whence the danger to his patients. Reckless stabs in the pelvis, by one unacquainted with the exact situation of the various organs, are very liable to be followed by peritonitis and death, and indeed it is generally in consequence of this result that such cases are brought to light.

Medicinal means are far more frequently used than violent measures, but it is doubtful whether they are often successful. Most of them are only likely to be so at the expense of the life of the woman. The minerals which are considered to have some effect in this direction are arsenic, corrosive sublimate, chromate of potash, and sulphate of iron, though probably none of them could produce abortion without killing the patient. Certain purgatives have enjoyed some reputation, amongst them are croton oil, elaterium, gamboge, aloes, and colocynth. There is no doubt that a powerful purgative, acting especially on the rectum, might excite the contractions of the uterus. Amongst other substances which have some reputation, and which, if they exert any influence at all on the uterus, must do it indirectly, may be mentioned pennyroyal, broom, fern, tansy, hellebore, yew, and laburnum. Cantharides also may be included in those acting indirectly, as it has been supposed that by the great urinary irritation it causes, uterine contractions might be provoked.

There remain two groups of substances, which have justly some reputation as abortives, viz. : the emmenagogues and the ecbolics. Of the former savin is the chief, the other substances in this class being aloes, gamboge, rue, madder, and black hellebore. They are reputed to excite or promote the menses. Ecbolics, on the other hand, act on the vascular system of the uterus, and excite uterine contractions, thereby tending to promote the expulsion of its contents. The ergot of rye stands almost alone in this class, though borax has been thought to possess some ecbolic action.

Savin.

Savin, administered either in the form of an oil, or simply the powdered tops, is perhaps the best known popular abortive. It appears, however, that when it does cause abortion it always does so indirectly, and often at the cost of the life of the woman, intense inflammation of the uterus, and neighbouring tissues and organs, being found after death. Sometimes the symptoms are those of a violent irritant of the alimentary canal, the vomit and intestinal contents being of a dark green colour, and having the odour of savin. A case has, however, been recorded in which the symptoms were such as might have been produced by strychnia, though there was no doubt that the death was due to savin. Taylor comments upon an important case where a medical man was convicted of administering oil of savin to a pregnant woman, with intent to procure abortion as follows :—It appeared that the prisoner had given the woman fourteen drops of the oil, divided into three doses daily, a quantity which was greater than should have been prescribed for any lawful purpose. His criminality rested not so much on the dose given, however, as on whether he knew, or as a medical man had reason to suspect, that the female, for whom he prescribed it, was pregnant. No medical authority would recommend oil of savin in full doses for pregnant women; and with regard to the existence or non-existence of pregnancy in special cases, medical men are reasonably supposed to have better means of satisfying themselves than non-professional persons. The prisoner's innocence therefore rested on the presumption, that he implicitly believed what the prosecutrix told him with regard to her condition, and that he had no reason to suspect her pregnancy, and therefore did not hesitate to select and prescribe a drug which certainly has an evil reputation, and is rarely used by regular practitioners. According to the evidence of the prosecutrix, she informed the prisoner that she had disease of the heart and liver, and that nothing more was the matter with her. It is absurd to suppose that oil of savin would be prescribed by a medical man for such a disease. The prisoner, on the hypothesis of innocence, must have intended the medicine to act on the uterus, and must have inferred the existence of an obstruction of menses from natural causes, irrespective of pregnancy. The jury do not appear to have given him credit for such ignorance of his profession, and he was convicted.

Ergot is a diseased growth occurring on the seed of rye, caused by a parasitic fungus. It acts powerfully on the uterus, and is largely used in uterine affections, both by the mouth and hypodermically. It produces its effects through the unstriped muscular coat of the arteries throughout the body, on which it has a powerful influence, and it has been recommended as a hæmostatic, in hæmoptysis and purpura. Its action, however, is a little uncertain, as was seen in the case of a woman with a deformed pelvis, to whom it was successfully administered in three consecutive pregnancies with a view to abortion, but totally failed at a fourth. Some writers have held that under its employment the mortality amongst the infants is greatly increased, but others have not found this to be the case. The explanation, if it were true, would probably be in the action exerted on the fœtal circulation. Large doses of ergot are not nearly so effective as small repeated ones. The action of the long-continued use of ergot is considered under the head of Poisons.

Ergot.

The signs of abortion are practically the same

Signs of abortion. as those of delivery at some period before the full term. If therefore, in the living woman, several days have passed since the abortion, it may be difficult if not impossible to prove the fact of its having taken place. The difficulties will be greater in proportion to the immaturity of the ovum. In the dead body, if too long an interval has not elapsed between the abortion and the death (which, however, is not usual if the death has any connection with the abortion), it will not be difficult to recognise the fact that, at no very distant date, there must have been an expulsion of an ovum from the uterus. Care must always be taken not to mistake the appearances due to menstruation for those due to abortion. When a woman dies whilst the catamenial period is upon her, the following will be the conditions found: The walls of the uterus will be thickened and spongy, and its mucous membrane more or less swollen and suffused, the os uteri somewhat patulous, and the ovaries will also be swollen. The value of the presence of corpora lutea has already been discussed (*vide* page 320), and it is unnecessary to add anything further on that subject.

With regard to the use of instruments to procure abortion it may be said that, except in skilled hands, they probably could not be used, so as to procure abortion, without leaving behind such marks as would easily be recognised. They all act by puncturing the membranes, but, unless the weapon is used with great care and skill, it is unlikely that the membranes would be ruptured without some damage being done to the cervix or the vagina. In all cases of abortion, therefore, where there is any reason to suspect that there may have been criminal violence, these parts should be most carefully examined. The absence of any lacerations, or signs of injury, about them would be considered as tending to show, that there was no evidence to support a charge of criminal abortion, by mechanical means at any rate. The most common cause of death in these cases is peritonitis, but it by no means follows, when this happens, that it is not due to other causes than violence.

Difficulty has sometimes arisen in cases of extra-uterine fœtation, where at first the symptoms are those of ordinary pregnancy, though in the later stages they are very different. In one such case the very sudden death of a young lady, known to be pregnant, shortly after taking some medicine gave rise to the suspicion that her death had, in some way, depended upon the medicine. Upon post-mortem examination, however, it was found that the case had been one of extra-uterine fœtation, and that the cyst containing the embryo had ruptured, permitting the escape of a large quantity of blood into the peritonæum, a condition which fully and satisfactorily accounted for her sudden death.

CHAPTER XXXII.

LEGITIMACY—DURATION OF PREGNANCY.

Duration of pregnancy.
EVERY child born in wedlock, or within a period, after the death of the husband, compatible with the ordinary duration of gestation, will be regarded by the law of England as legitimate, unless the contrary is proved by evidence. Medical evidence, in questions of this kind, is not of so much value as moral evidence. The former is only offered to show that the husband was so disabled by disease, or congenital defect, that he could not have been the father, an opinion, however, which should always be given with caution. If it can be shown that the husband could not possibly have had access to his wife, within a period not exceeding the limits of gestation (what the law is as regards those limits will shortly be shown), the child so born will be accounted illegitimate. There is a difference between the law of England and that of Scotland, inasmuch as in the latter country marriage legitimises children born before the marriage. A child that is born in lawful wedlock is assumed to be legitimate, and therefore if a man marries a woman who is known to be pregnant, or whom he ought to know to be pregnant, the child born will be regarded as legitimate, although it may be certain that the husband was not the father of it; the mere fact of his marrying the woman, when in that

state, is considered to show that he was willing to assume the responsibility of her condition.

The English law does not define the duration of the natural period of gestation. The great majority of children are born between the thirty-eighth and fortieth weeks after conception. Many persons believe, that the duration of pregnancy is really a multiple of the menstrual period, and that delivery under ordinary circumstances takes place at what would have been the tenth return of the menstrual period, and the fact that it is said that abortion is more likely to occur at the date of a menstrual period than at other times affords support to this hypothesis. Some have suggested that cases of prolonged gestation were apparent rather than real, being in truth cases of protracted parturition, but there is very little evidence in support of this view, or of the view that gestation is longer in the case of male children than female.

It is impossible to determine the exact date at which conception takes place; there is no reason to believe that it necessarily takes place at the time of, or on the same day as, coition. Impregnation is more likely to take place just after a menstrual period than at any other time, the next most likely time being immediately before menstruation, the period midway between these two epochs being the least favourable for impregnation. There is, however, no time at which it may not take place. In the absence of accurate information as to the date of conception, it is customary to commence the calculation from the date of the last appearance of the menses.

All children born before the thirty-eighth week may be considered as premature, though it by no means follows that they will not live, or that it would be possible for a medical witness to swear that the child had not been born at the full time.

Viability.

It seems probable that children do not grow much after a certain time in the uterus, for, in undoubted cases of protracted gestation, the children are not bigger than those born at the full time. The earliest age at which a child can be born alive, and survive its birth, is often a matter of some importance in determining the legitimacy of a child, and many important trials have taken place on this question of the viability of children. It is a question more often raised in France than in England, for in the former country a child is not considered viable, even though it has lived for a day or so, if, after death, some congenital abnormality is found incompatible with an existence of more than a few hours' duration. Under such circumstances a child would be pronounced non-viable, and therefore incapable of inheriting. In England the viability of a child is a matter for direct evidence; if it can be shown that it had any signs of life after being separated from its mother, the mere fact will suffice to enable it to inherit and transmit property.

Illustrative cases.

The following are instances of unusually short duration of pregnancy. In the first (Dr. Outrepont's) there was the strongest reason to believe that gestation could not have exceeded 27 weeks; the child weighed $1\frac{1}{2}$ lb. at birth, and measured $13\frac{1}{2}$ inches. The skin was covered with down, and much wrinkled; the limbs were small, the nails appeared like white folds of skin, the testicles had not descended. It breathed as soon as it was born, and by very great care its life was saved. Until it had reached a time corresponding with the 42nd week of utero-gestation its development was slow. At the age of eleven years the child was about the size, and had the appearance, of a child of eight years old. In the other case (Dr. Halpin's) a healthy woman, the mother of five children, was delivered in the sixth month of her pregnancy of a female child; contrary to expectation, the child survived, and was healthy in every respect. The weight of the child, on the fourth day, after birth, was 2 lbs. 13 oz., and on the 34th day 3 lbs. 8 oz. Four months after birth she weighed 8 lbs. 8 oz. In each instance the weight of the child confirms the probable date of the gestation. Children born before the completion of the sixth month of gestation rarely live more than a few hours; if born after that period they may, with very great care, survive.

The signs of maturity in a new-born infant will be considered in detail under the head of infanticide; a few points may be briefly alluded to here. It is a well-known fact that the size and weight of a newly-born child, at the full term, may vary within wide limits; the majority are probably between ·6 and 8 lbs. in weight, and yet it is by no means uncommon to find a child weighing 10, 12, or even 14 lbs. at birth. In such cases it is presumable that at the eighth month the child would be larger, and weigh more, than one which was only going to weigh 6 lbs. at the full time; it is therefore highly probable that an eight months', or even a seven months', child might weigh as much as one born at the full time. Because a child weighs, say, 7 lbs., it does not follow therefore that it is of full time. If any fœtal peculiarities remained, such as the membranæ pupillares, or if in the male the testicles had not descended into the scrotum, or if there were other obvious signs of immaturity, it might be possible to speak with certainty to the child's not having reached the full term. It is hardly conceivable that a child at the sixth month of gestation should present all the characters of maturity. *Maturity.*

As a rule, in the same woman no two pregnancies have exactly the same duration. It must *Limits of the duration of pregnancy.*

be admitted as a possible explanation of very protracted cases, that the menses may have ceased from some totally different cause, before conception took place, and that therefore it is incorrect to date the conception from the time of their disappearance; and, indeed, it has been shown that conception may take place when the catamenia are entirely absent. The best answer to this objection would be, that some cause, other than pregnancy, should be shown why the catamenia should have stopped at that particular time, if the objection is to be allowed much weight. The late Dr. Reid drew up the following propositions, in reference to the duration of pregnancy: '1. The duration of pregnancy is not altogether a fixed period; it varies somewhat in the human female. 2. This deviation is not, however, to any great extent; the only certain data of calculation are those dependent on the known time of conception. 3. The average duration of the pregnant state is about 275 days, or between 270 and 280 days. 4. There is no full or satisfactory evidence that gestation has been prolonged beyond 293 days. The French Code, which allows 300 days, may be regarded as liberal. 5. The menstrual period must generally serve as our guide in default of some exact knowledge; it is, however, often fallacious, and only a means of approximation to the probable time of parturition. 6. The fortieth week after the last appearance of the menses is the most likely period, the forty-first week the next.' The French law, as has just been stated, allows 300 days, whilst the Prussian law gives 301, as the extreme limit for gestation; a child born after a longer period than that, in either of those countries, will be, *ipso facto*, illegitimate. The English law recognises no fixed date, and the legitimacy, in any disputed case, will be settled by the evidence, medical and moral, and not in accordance with any dogmatic rule. The possibility of a gestation lasting 317 days was affirmed by a jury in one case.

The Gardner Peerage case is probably one of the most important in which the question of legitimacy has been raised in this country. Allan Legge Gardner, the son of Lord Gardner by his second wife, petitioned to have his name inscribed as a peer on the Parliament Roll. The peerage was, however, claimed by Henry Fenton Jadis, *alias* Gardner, who said that he was the son of Lord Gardner by his first, and subsequently divorced, wife. It was contended that the latter claimant was illegitimate, and in order to establish this point the evidence was partly medical and partly moral. Lady Gardner, the mother of the alleged illegitimate son, parted from her husband on January 30, the latter going to the West Indies, where he remained some time, and did not see his wife again till July 11. The child in question was born on December 8. If the child was legitimate, therefore, he was born after a minimum gestation of 311 days, or a maximum one of 150 days. The latter hypothesis was not entertained at all, as the child was undoubtedly not premature when born, and the only medical question was, whether so long a period of gestation was probable. The majority of medical witnesses agreed in thinking that such a duration of pregnancy was possible, and that, on medical grounds alone, there was no reason for declaring the child illegitimate. The case was decided against Jadis, but mainly on moral grounds.

The Gardner Peerage case.

It sometimes happens that conception occurs in a woman who is already pregnant, gestation proceeding independently in each case, so that she is delivered of two full-grown children at different dates. In animals there can be no doubt of the

Superfœtation.

possibility of such an occurrence, and the fact, that there have been several instances of women giving birth to twins of different colours, would seem to be conclusive of the possibility of super-conception. The older writers, however, including Casper, do not admit that a second conception can take place, except at a very short interval after the first, and they regard the closure of the uterus, by the development of the ovum, as an absolute bar to a second conception. All admit the possibility that in the case of twins one may be born a little before the full term, and the other at, or even a little after, the full term, and in this way an interval of several weeks may be obtained between the births of successive viable children. It seems possible also that in some cases of twin pregnancy, one twin may so press upon the other as to interfere with its due development, and, not being expelled at the full term with the first, it may remain in the uterus some little time longer, until it too has reached the full development; in this way two mature children might perhaps be born at intervals of a month. In those rare cases where the uterus is double there is less difficulty in admitting the possibility of superfœtation, especially when the vagina is also double.

When the interval between the births of successive mature, or at any rate viable, children amounts to four months or nearly so, none of the hypotheses which have been mentioned above can be said to offer a satisfactory explanation, and the majority of recent writers in this country agree in accepting the possibility of superfœtation, as the only means of explaining such cases. The following cases are generally regarded as instances of superfœtation:—The first concerns a lady who had fourteen children. Her fourth was born on July 29, 1782, and lived to the age of 60 years. The next to this was born on January 19, 1783, and reached the age of 29. Only 174 days had elapsed between the two deliveries, and, supposing conception to have taken place after the birth of the first of them, it would be necessary to admit that a child born after 167 days of intra-uterine life could reach maturity, which is quite contrary to all evidence on the subject. In the other case a lady gave birth to a child on September 19, 1849, and another on January 24, 1850, both of whom reached maturity. The interval between these deliveries was 127 days, much too short an interval for the gestation of a child that was mature, or rather viable, at birth.

In the case of monsters it will probably be the best plan to adopt Dr. Tidy's advice, and leave the law always to decide whether, in any given case, the deformity is such as to amount to monstrosity. According to Coke, a monster is a being which hath not human shape, and cannot therefore inherit property. It is obvious therefore that the question of whether a being is a monster or not must, sooner or later, be decided by the law, and it will be well for medical men to abstain from expressing an opinion on the point. Monsters are of several kinds: when there is no head they are called acephalous; when there are two heads with one body, dicephalous; and when there are two bodies with one head, disomatous. The general rule hitherto has been to consider a monster with two heads as two beings, and when there is only one head, as one being. It is said that in Paris, in the seventeenth century, a two-headed being was sentenced to death for stabbing a man, but was not executed—on these grounds.

A very remarkable instance of monstrosity occurred in the case of the Siamese twins. These men, who lived to be over sixty years of age, were really two separate beings united by a broad band

Monsters.

opposite the ensiform cartilage. All their viscera were perfectly distinct, as also were their circulatory systems and their mental faculties; indeed, they were of very different temperaments. In all probability the band might have been divided with perfect safety during life, as it was ascertained after death that although a prolongation of peritonæum entered it from each side, these did not communicate, but each terminated in a *cul-de-sac*. The twin females known as Miss Millie Christine are more intimately united, the union being at the back, commencing at the second lumbar vertebra, and extending down to the coccyx. They have four arms and four legs, and as regards their intellectual capacities are perfectly distinct beings.

Plural births.
Twins are, of course, not at all uncommon, and triplets are not so very rare; and, if ancient writers may be believed, women have been known to bring forth four, and even five, children at a birth, the children all surviving. The only importance attaching to these cases is to determine which of the children was born first.

Supposititious children.
Many reasons may combine to make a woman endeavour to represent the children of another as her own, and it is hardly necessary to enter here into the consideration of such motives. In the most recent case in this country, it was shown beyond doubt that the woman was insane.

Disputed paternity.
Questions in reference to paternity, as a rule, only arise in connection with the subject of bastardy, and then the evidence does not carry much weight; nor would a general similarity of features; but if there is added a similarity in manner, voice, and demeanour, more stress might be laid upon it. Almost the only case in which the question of identity has turned upon likeness was that known as the Douglas case. Archibald Douglas, the survivor of two brothers, claimed the title after the

death of his father, Sir J. Douglas. Much importance, in favour of the legitimacy of these children, was attached to the fact that they resembled, the one Sir John, the other Lady Douglas. Lord Mansfield, in delivering judgment, said : ' I have always considered likeness as an argument of a child being the son of a parent, and the rather as the distinction between individuals in the human species is more discernible than between other animals. A man may survey ten thousand people before he sees two faces exactly alike; and in an army of ten thousand men, every man may be known from another. If there should be a likeness of feature there may be a difference in the voice, gestures, or other characters, whereas a family likeness runs through all of these, for in everything there is a resemblance as of feature, voice, attitude, and action.' The opposite doctrine, however, could not be allowed, viz. that a child should be accounted illegitimate, because he did not resemble his parent.

There are other grounds, however, on which the legitimacy of a child may be called in question, besides those already considered, viz. those based on alleged, or real, sexual incapacity on the part of one or other parent. These may be due to abnormal causes, *e.g.* congenital defect or the results of disease, or to normal conditions, *e.g.* natural decay or immaturity. Hermaphroditism is the name given to certain forms of congenital arrest of development of the genital organs, whether complete or incomplete. The word, in its strict and original meaning, was intended for those cases in which there is an admixture of both sexes, but it has long since been applied to almost any form of congenital defect. In the fœtus, until after the fourth month, the genito-urinary organs are not sufficiently differentiated to enable an observer to

Hermaphroditism.

determine to which sex the child would have belonged. One of the most common forms of male hermaphrodite is that in which the two halves of the scrotum have not united, so that an apparent vulva is produced, but there is no vagina; the penis may be well formed, and the opening of the urethra may be either in the normal place, or on the under surface of the penis, a condition known as hypospadias. In such a case the testicles would be found one on either side in the apparent vulva, and there is no necessary reason why they should not be perfectly well developed, and the person capable of begetting children. The female type of hermaphrodite consists chiefly in an unusual development of the clitoris, so that it resembles a penis except in not being traversed by the urethra; in these cases the vagina is generally more or less imperforate, and the uterus and ovaries either absent or undeveloped. It is important in all these cases that a decision should be arrived at, if possible, as to which sex the being belongs. If testicles are present, the male sex must be allowed to predominate; if a vagina and ovaries, then the female sex. The difficulty lies in determining the absence of the testicles, as these may be undescended, though well developed. The general conformation, especially as regards the pelvis and shoulders, should be taken into consideration, and, to a less extent, as regards the presence of hair on the face, and the character of the voice. In some cases it is impossible to assign the being to either sex; when, for instance, there are neither testicles nor ovaries, the being has no sex at all. On the other hand, sometimes there is an admixture of both sexes, as in the case of a being who died at the age of fifty-five, and concerning whose sex, during life, there had been much speculation and difference of opinion. On post-mortem examination there was found on the right side a withered testicle with a penis and prostate gland, as male peculiarities; whilst on the left side there was an ovary with a uterus, vagina, and Fallopian tube, as female peculiarities. The general configuration of the body was that of a woman.

CHAPTER XXXIII.

IMPOTENCE, STERILITY, RAPE, UNNATURAL OFFENCES.

Impotence and Sterility.

Causes in the male: Congenital.

In the male. The causes of these conditions may be divided into congenital, natural, and acquired. The *congenital* causes include those conditions which have just been under consideration in the various forms of hermaphroditism. If the penis is absent, the person must be considered impotent, even if the testicles are well developed. An ordinary case of epispadias or hypospadias does not necessarily imply impotence, but if the opening of the urethra is so situated that during copulation the emission of semen into any part of the vagina is impossible, then such a condition would amount to impotence. Over-development of the penis could not *per se* be regarded as a source of impotence. The mere non-descent of one testicle does not afford any evidence of impotence, as the other one may be quite sound, and even the undescended one, though often somewhat ill-developed, is not by any means to be regarded as functionally useless, for it is certain that a person, both of whose testicles are undescended, may yet be capable of fruitful intercourse. When both testicles are undescended, it is difficult, as has already been observed, to determine whether they are present or absent, supposing that neither can be felt in the inguinal canal. If the semen, on repeated examination, should not be found to contain any spermatozoa, there would be a strong presumption that the testicles were either altogether absent, or present only in a rudimentary condition. A single examination for spermatozoa, if negative, should be confirmed by subsequent examinations, before an opinion is given. When, in the absence of any external signs of the testicles, no proof of the presence of spermatozoa can be obtained, the person must be regarded as incurably sterile.

The *natural* causes of impotence are juvenility and senility. Before the arrival of puberty procreation is impossible: the law, however, does not lay down any precise date for this period, but from the fact that fourteen is the age fixed upon as that at which a boy may marry, it is to be presumed that it regards him as incapable of procreation before that age; and a boy under the age of fourteen has been decided to be incapable of committing a rape. The presence of the general signs of virility would be useful in deciding, whether any given lad had reached the period of puberty or not. There is no law as to the date at which the procreative power may cease, and, on more than one occasion, the judges, in this country, have refused to allow the age of a man to be used as an argument in support of his impotence. The only evidence that would be of any value would be that of the absence of spermatozoa, and Casper mentions having found them in the semen in a man of ninety. <small>Natural.</small>

Amongst the *acquired* causes castration naturally occupies the first place; when both testicles have been completely removed, such a person must be ever afterwards absolutely sterile, though there is reason to believe that fruitful intercourse may take place within a limited period of the operation, owing to the presence in the vesiculæ seminales of <small>Acquired.</small>

some semen. The removal of one testicle for disease is not followed by sterility, though the other testicle may sometimes atrophy under these circumstances. Syphilitic, or strumous, disease of the testicles leads sooner or later to sterility. Next to local diseases, those affecting the spinal cord are apt to be followed by impotence, especially locomotor ataxy, so much so that impaired sexual power is ranked as one of the symptoms of that disorder. Disease confined to the lateral columns of the spinal cord is not associated with impotence Diseases of the cerebellum, and of the brain, are generally followed by impotence, but not always; at any rate they are not considered to be so by the law, for in the case of Legge *versus* Edmunds, a few years ago, a child was held to be legitimate which was born four months after the death of the father, the parents having been married eight years, and this being the first child. Moreover, the father, who had been a very dissipated man, had had an attack of hemiplegia seven months before his death, which lasted up to his death; and further, the mother married again a year later, and subsequently had four children, and there was some ground for suspicion that she had been unduly intimate with this second husband, during the lifetime of the first.

Any general disease attended with great debility will be associated with impotence, which may be only transitory; in diabetes it is often a marked symptom, and, among the acute diseases, mumps may be mentioned as one, which wasting of the testicles and incurable sterility have, according to the older writers, been known to follow. Excessive indulgence in alcohol, opium, or tobacco, or the long-continued use of bromide of potassium, may be a cause of impotence; sometimes, too, it has supervened after a blow on the back of the head.

In the female. Here too the same classification may be adopted. Amongst the *congenital* causes, those in which there is entire absence of uterus and ovaries are of course irremediable, but when the defect merely consists in a more or less complete atresia of the vagina, the barrenness may be entirely removed by an operation. It is possible that if the uterus were undeveloped, but the ovaries were sufficiently developed, one of the varieties of extra-uterine pregnancy might take place. If the barrenness is due to some incurable congenital defect, and if the husband had no reason to know of the existence of such a defect at the time of marriage, then such a marriage is voidable, —not necessarily void, but voidable if the husband take the proper legal steps; no third party is allowed to interfere in such a case. The same holds good in the case of incurable congenital impotency in the husband; the wife may get the marriage declared void, but no one else can do so for her.

Causes in the female: Congenital.

As a rule the establishment of menstruation marks the arrival of the generative period in woman, and prior to puberty she is not capable of conceiving. There have, however, been undoubted instances of conception by women who had never menstruated, and women have been known to pass through the whole of the child-bearing epoch, after once becoming pregnant, without again menstruating, and have yet during that period borne several children. The law allows a girl to marry at the age of twelve, but in this country puberty is not, as a rule, reached before the age of fifteen. A case has been recorded of a girl who at the age of one year commenced to menstruate, and at the age of ten gave birth to a child. The catamenia usually cease between forty-five and fifty years of age, and their disappearance usually coincides with the cessation of the child-bearing power; one instance has occurred, however, of the birth of children after the final disappearance of the menses, though no doubt such cases are rare. Few women bear children after attaining the age of fifty, but cases have been put on record in which women have borne children at the ages of sixty-five and seventy-three respectively. For a woman to commence child-bearing after the age of fifty, must be regarded as quite exceptional. Amongst the acquired causes hysterectomy, or removal of the uterus (Porro's operation as it is called when it is performed on a woman after delivery of the child by the Cæsarian section), is of course followed by sterility, as also is the removal of both ovaries, an operation which, of late years, has been deprived of much of its formidable character by the advances in abdominal surgery. The removal of one ovary, as in an ordinary case of ovariotomy, is not necessarily followed by sterility, any more than is the loss of one testicle in the male.

Spencer Wells on some causes of sterility.

In a paper on 'Some Remediable Causes of Sterility in the Female,' Sir Spencer Wells gives the following summary: ' 1. The ovaries not performing their normal function of periodical ovulation; in other words, an absence of ova prepared for impregnation. 2. Alterations in the coverings of the ovary, interfering with the escape of the ovum. 3. The ovum escaping from the ovary, but failing to be grasped by the fimbriæ. In such case either the ovum perishes, or, if it be impregnated, extrauterine fœtation results. 4. Alterations in the Fallopian tubes, obstructing the passage of the ovum. 5. Alterations in the uterus or vagina, or in the external organs, preventing impregnation of the ovum either by preventing access of the spermatozoa to the ovum, or by the destructive action of unhealthy fluids on the spermatozoa, or from some other interference with the physical conditions of fecundation.' This includes such various states as occlusion of the vagina by adhesion of the labia, or imperforate hymen, eczema, or herpes of the vulva, hyperæsthesia and anæsthesia of the vagina, diseases of the rectum, urethra, or bladder, ruptured perinæum, vaginal and uterine leucorrhœa, diseases and displacements of the uterus, contraction of the os and cervical canal, foreign bodies and tumours in the uterine cavity, including cancers, polypi, fibroids, &c.

As a general rule it may be said that any inflammatory disease of the uterus, or ovaries, may be followed by sterility. It is a well-known fact, for which no explanation has as yet been forthcoming, that a woman who is sterile with one husband may prove fertile with another, and this when there is no reason to suppose sterility on the part of the husband.

Dr. Matthews Duncan entered at some length into the subject of sterility in the female in his Gulstonian Lectures for 1883. His definition of sterility is 'the inability to produce living and viable children.' It may be absolute, not absolute, and relative. Absolute sterility is where no conception takes place, not absolute where conception does occur, but the ovum is cast off before reaching maturity. Relative sterility includes all cases where the number of children, produced by a woman, is under the average number produced by women under similar circumstances. It may occur either as an exhaustion of the reproductive power alone, or with failure of constitutional strength as well. A woman who does not bear a child within sixteen months of marriage shows some degree of sterility, as also do those who do not bear children about every twenty months of their married life during the child-bearing period. Amongst the

Matthews Duncan's classification.

constitutional causes he gives age : before twenty, and after twenty-four, there is a degree of comparative sterility; early marriage favours sterility. The two pathological causes, upon which he lays most stress, are dysmenorrhœa and functional diseases of the generative apparatus, leading to, or accompanied by, absence of sexual desire and pleasure. Dysmenorrhœa he defines to be a neurosis, characterised by painful spasms of the uterus at the menstrual period, independently of any gross structural changes in the uterus itself. The absence of sexual desire and pleasure is also to be regarded as a neurosis; excessive sexual desire he regards as an evidence of deficient sexual power.

Rape.

Definition. The legal definition of rape is the carnal knowledge of a woman by force and against her will. By carnal knowledge is understood simply penetration of the male organ within the vulva, and not necessarily into the vagina; seminal emission is not necessary, and the crime may be quite complete without this. The following is the statute relating to the crime of rape: 'Whosoever shall be convicted of the crime of rape shall be guilty of felony, and, being convicted thereof, shall be liable, at the discretion of the Court, to be kept in penal servitude for life, or for any term not less than three years, or to be imprisoned for any term not exceeding two years, with, or without, hard labour.'

Question of consent. In the case of a woman, if she consents the crime is not rape, unless such consent had been extracted from her under fear of death. It has often been doubted whether a rape could be perpetrated on a woman of ordinary strength and in good health, but the paralysing influence of terror must in some cases be allowed to have some force.

It is not to be believed that a rape could be committed on a virgin in a state of ordinary sleep, though such a thing may be possible in the case of a woman accustomed to sexual intercourse. Rape may be committed on a person whilst under the influence of narcotics, alcohol, or anæsthetics, though all cases in which rape is said to have been thus committed should be received with suspicion, more especially if, in the case of an anæsthetic, the woman represents its effects as having been sudden. A woman who is an imbecile or idiot cannot give her consent, and therefore rape on such a person is a felony.

As regards children, the statutes run : 'Whosoever shall unlawfully and carnally know and abuse any girl under the age of twelve years shall be guilty of felony,' and ' Whosoever shall unlawfully and carnally know and abuse any girl being above the age of twelve years and under the age of thirteen years, whether with or without her consent, shall be guilty of a misdemeanour.' This is not interpreted to mean that a rape on a girl under thirteen is not a felony. A boy under the age of fourteen cannot be charged with rape, though in a great many cases he would be able to commit the crime. Rape of children.

Medical evidence on the question of rape is only available when the charge is made soon after the commission of the alleged deed, and the points on which it would bear would be the swelling and signs of inflammation about the vulva, rupture or laceration of the fourchette, or hymen, the presence of blood or dried semen about the vagina, vulva, or labia, and, in the case of a woman over the age of thirteen, marks about the body indicative that there had been resistance offered, and that a struggle had taken place. Since, however, proof of vulval penetration only is required, the signs of Signs of rape.

A A

violence about the genitals may be entirely absent, and yet a rape may have been perpetrated. The presence of gonorrhœa, or syphilis, in both the accuser and the accused, would be corroborative evidence of the truth of the charge, but it is difficult to distinguish between gonorrhœa and simple leucorrhœa. The presence of disease in the accuser and not in the accused would go far to show that the case was not *bonâ fide*.

It is in the case of children that charges of rape are most common. A mother discovers that her little girl has some discharge from the vagina, and forthwith concludes that the child has been violated; in some instances, it is only too certain that the child has been made to bring a perfectly groundless charge against someone whom her parents wished to spite. Such instances have occurred in our own law courts, and they would seem to be even more common in other countries. A slight vaginal discharge is quite common in young and delicate children, almost as common as otorrhœa, and may come on without traumatic cause of any kind; it may or may not be associated with redness and swelling of the labia; in bad cases there may be serious inflammation of the vulva. Sometimes its existence coincides with the presence of threadworms about the anus, but more often no source of local irritation can be found. In the vast majority of cases, charges of rape are probably false. It has been suggested that pregnancy cannot follow rape, but there is no reason why it should not. The possibility of rape being the actual cause of death has been placed beyond doubt by the well-known case of Amos Greenwood.

Unnatural Offences.

Under this head are included sodomy, or unnatural intercourse between man and man, or between man and woman, pæderasty (the term applied to sodomy when the passive agent is a boy), and bestiality, which implies intercourse between man and one of the lower animals. The punishment for any of these offences is penal servitude for life, or for any term not less than ten years.

Both parties to an unnatural offence are punished by the law, consent being in every case presumed to be given, as without it the commission of such a crime would be impossible; but when the intercourse is thus practised on a person under the age of fourteen years, such a person would not be tried, the active agent alone being charged, it being presumed that the other was of too tender an age to understand the nature of the act.

CHAPTER XXXIV.

BIRTH AND INHERITANCE.

Birth.

What constitutes live birth. ONE of the most important and disputed points in medico-legal matters, is that which relates to the live birth of children. Legally, a child is not considered to have been born alive unless it has manifested some undoubted sign of life after its entire birth; though separation from its mother is not necessary, the child must have had a separate existence after it came into this world, if it is to be regarded as having been born alive. Evidence that a child had breathed, or cried, after its entire birth would, of course, be conclusive of live birth, but it is not necessary to prove either of these in order to establish the fact of live birth. The recognition of the heart's pulsation, or even the spasmodic twitching of a muscle, would be sufficient evidence of live birth, if clearly proved. Nor is it necessary to prove any continuance of these movements, for, unless it be admitted that the heart of a dead child can pulsate, or the muscle of a dead child twitch, the mere fact of the single occurrence of either of these will be a proof, that at that moment the child was actually alive. As will shortly be seen, in the case of infanticide, the law requires much more stringent proof of live birth than it does in respect of civil cases; indeed, in regard to trials for infanticide, the attitude taken by the law is to assume that the child was still-born, unless the contrary is shown.

The following case affords a good instance of the kind of evidence that is accepted in these cases. A married woman died, and, in default of issue, her estate went to her next-of-kin. Some years later, however, her husband learnt that there was evidence that a child, to which his wife had given birth, had not been still-born, as he had always supposed, but had been born alive. He then brought an action to recover the estate, as, if his wife had given birth to a living child during their married life, he was entitled to become tenant for life of the estate. The accoucheur who had been present at the confinement was dead, but it was proved that he had declared the child to have been living an hour before its birth, that he had directed a warm bath to be given to it, and that when the child was born, he gave it to the nurse to place in the bath. The child neither cried nor moved after its birth, nor did it manifest any active sign of its existence, but the two women, who placed it in the bath, deposed that they twice noticed a tremulous motion about the lips. They informed the accoucheur, who directed them to blow into the child's throat, but it did not show any further signs of life. The main question at the trial was whether this tremulous motion of the lips was a sufficient sign of live birth. The argument in support of this contention was that, if the child had really been born dead, there could have been no muscular movement, however slight. The Court decided that the child had been born alive.

It is well known that a child may live for some time after birth, and not breathe, the lungs for some cause or other remaining in their fœtal state of atelectasis; but, in such a case, life may be most clearly manifested by the pulsations of the heart, and if these are recognised there could be no denying the fact of life.

<small>Crying not a fair test.</small>

The Scottish law requires that a child must have cried after its entire birth, before it is considered to have been born alive. This is obviously an unfair and unreliable test, for it is perfectly well known that a child may live for some time without making any sound, though no doubt the majority do cry very soon after birth. Coke quaintly suggests as an objection to this sign that the child may be born dumb, and so unable to cry. A more important point to be borne in mind is that the child may cry before it is fully born, when the head only is extruded, the body remaining in the vagina, and, after so crying, it may die before its entire birth. So that practically the evidence of an eye-witness is necessary to establish the fact of live birth.

Inheritance.

According to the English law, a child that is born alive can inherit and transmit property to its heirs, even if it did not live a minute, and even if its death was due to some congenital defect, incompatible with life. All that the law requires is that the child should have had, no matter for how short a time, a separate existence in this world. In France there is a material difference in regard to the law on this subject, for there it is necessary, not only that a child should be born alive, but further, that it should be viable, *i.e.* that it should be capable of living, and if the child die within a few hours of its birth, and it is found that its death was due to some congenital anomaly, inconsistent with the due performance of any vital function, such an infant is liable to be pronounced non-viable, and thereby declared to be incapable of inheriting.

INHERITANCE. 359

Reference has already been made to tenancy by courtesy. Briefly, what is meant by this phrase is as follows: When a married woman, possessed of some property, dies before her husband, and leaves no children, her property passes at once to her heirs-at-law, unless she shall have given birth to a living child during her married life; in this case her husband is allowed by law to become a tenant by courtesy for life of her estate. What would happen if the child were born alive after the death of its mother, as, for instance, by the Cæsarian section, it is impossible to say; the presumption would be at first that, as the child could not be said to have been born during the lifetime of its parents, the father could not become a tenant by courtesy. It follows from what has already been said about monsters, that the child born must be capable of inheriting, that is, it must not be a monster, if the father is to have this right granted him.

Tenancy by courtesy.

Even before its birth, and whilst still *in utero*, the fœtus has certain legal rights. An estate or money may be left to it, and a guardian may be assigned to it, though of course these are dependent upon its being born alive. The fœtus *in utero* can be appointed an executor, but the child cannot act as such until he has attained the age of seventeen years.

Legal relationships of age.

The exact date of birth is sometimes of legal importance, and it must be taken from the moment of complete birth, and not from the time when the head was first extruded, or when the child cried. In law, a person is considered to be of age on the day preceding the twenty-first anniversary of his birth, and since, in law, a part of a day counts as a whole day, he is considered to be of age from the first minute of that day, though it might be nearly forty-eight hours short of the real anniversary of his birth, supposing him to have been born, that is to say, just before midnight. After the age of twenty-one all legal disabilities are removed. A child under the age of seven years is presumed to be incapable of committing a crime; after that and before fourteen years there is no presumption of law, and it will be a matter to be proved by evidence, whether he knew that he was doing wrong when he committed a crime; after the age of fourteen a person is fully responsible for his actions. In reference to the marriage of people under the age of twenty-one, the law requires the consent of the nearest relative or guardian, except in the case of a person who has already been married.

CHAPTER XXXV.

INFANTICIDE.

INFANTICIDE means the murder of a newly-born child, and may be taken to include that of a child a few days old. As has been stated in a previous chapter, the law assumes in the case of a child found dead that it was still-born, and therefore, before proving that a murder has been committed, it is necessary to prove that the child was entirely born, and living, at the time that the murderous violence was inflicted. In many instances this is impossible.

Since a natural death is more probable in the case of an immature than a mature child, it is necessary that the signs of maturity should be well known. The English law, as has been already mentioned, does not concern itself with the viability of children, but, medically speaking, proof of the immaturity of a child would afford a presumption either of its having been still-born, or having died from natural causes.

The following are the signs by which the intra-uterine age of a newly-born child may be inferred, as given by Tidy :—At *six* months: length, 9 to 10 inches; weight, 1 lb.; skin presents some appearance of fibrous structure; eyelids still agglutinated, and membrana pupillaris remains; sacculi begin to appear in the colon; funis is inserted a

Signs of maturity.

little above pubis; face of a purplish red; hair white or silvery; sebaceous covering (vernix caseosa) begins to present itself; meconium in large intestine; liver dark red, gall bladder contains serous fluid; testes near kidneys; points of ossification in four divisions of the sternum; middle point of body at lower end of sternum. At *seven* months: length, 13 to 15 inches; weight, 3 to 4 lbs.; skin rosy, thick, and fibrous; sebaceous covering appears; nails do not yet reach extremities of fingers; eyelids no longer adherent, membrana pupillaris disappearing; a point of ossification in the astragalus; meconium occupies nearly the whole of the large intestine, valvulæ conniventes appearing; cæcum in right iliac fossa; left lobe of liver nearly as large as right, gall bladder contains bile; brain firmer; testicles further off kidneys; middle point of body a little below end of sternum. At *eight* months: length, 14 to 16 inches; weight, 4 to 5 lbs.; vernix caseosa all over skin; nails reach extremities of fingers; membrana pupillaris becomes invisible during the month; a point of ossification in the last vertebra of sacrum; no centre of ossification yet in the cartilage of the lower extremity of the femur; convolutions begin to appear in brain; testicles descend into internal ring; middle point of body nearer umbilicus than sternum. At *nine* months: length, 17 to 21 inches; weight, 5 to 9 lbs.; head more or less covered with hair; skin, especially at heads of joints, still covered with sebaceous matter; membrana pupillaris gone; external auditory meatus still cartilaginous; four portions of occipital bone still remain distinct; os hyoides not ossified yet; point of ossification in the centre of cartilage at lower extremity of femur; white and grey matter of brain become distinct; liver reaches umbilicus; the testes have passed the internal ring, and are often found in the scrotum; meconium in rectum or sigmoid flexure; middle point of body at the umbilicus or a little below it.

The relative position at which the umbilicus is inserted into the body does not afford any reliable guide. It must not be forgotten that in the case of twins the children are usually smaller, and less well-developed, at birth than in the case of ordinary children; thus 5 lbs. would be a common weight for a twin child at the full time. It is difficult to speak positively as to the intra-uterine age of a child in any given case, but there should be no difficulty, as a rule, in determining that a seven months' child had not reached the full term of utero-gestation. There is one sign, by which the maturity of the child may sometimes be estimated with great nicety, upon which Casper lays some stress, a sign, however, only available after the death of the child. It is the presence of the osseous nucleus at the lower end of the femur, to which reference has already been made. It should be looked for, by carefully slicing off thin horizontal sections of the lower end of the femur until it is reached, when its size should be carefully measured. Casper's conclusions regarding it are the following: '1. When there is no visible trace of the centre of ossification in the inferior femoral epiphysis, then the fœtus can be no more than from thirty-six to thirty-seven weeks old. 2. The commencement of this osseous nucleus, which is at first about the size of a hemp-seed, indicates a fœtal age of from thirty-seven to thirty-eight weeks, supposing the child to have been still-born. In the opposite case, the child may have been born alive before the time, without any osseous nucleus, which then becomes developed during its extra-uterine life. In rare instances of unusually retarded development, a fœtus of forty weeks may exhibit only a trifling commencement

The nucleus at lower end of femur.

of this nucleus. When this osseous nucleus possesses a diameter of from three-quarters of a line to three lines, it indicates that the fœtus must have attained a uterine age of forty weeks, always supposing, of course, that the child has been still-born. In one instance of unusually retarded development, with defective ossification of the skull, of a girl born perfectly mature, we found no osseous nucleus. 4. We may conclude that the child has lived after its birth, when the osseous nucleus measures more than three lines. But, on the other hand, an osseous nucleus of less than three lines does not prove that the child has not lived.'

The best and most conclusive proof of stillbirth is evidence to show that the child died *in utero*. This cannot be obtained when the child died only a few hours before birth, as the appearances then do not differ from those of a child which has died at, or just after, its birth. Where the child has died, and remained dead for some days, in the uterus, it undergoes certain changes, due to the prolonged maceration in a warm fluid, which cannot be mistaken for the ordinary changes of putrefaction. The body of the child after maceration has a peculiar and penetrating odour; the skin is peeling in patches, and is of a coppery red colour, there being no trace of green in it; the surface where the skin has peeled off is moist and greasy; the bones are more or less separated; and there is a general flabbiness of the body, causing it to lose its shape, even after only ten days or so of maceration. When the body has remained longer in the uterus, it may become adipocerous, or encrusted with phosphate of lime. Ultimately, in those rare cases of missed labour where the dead fœtus is retained in the uterus for a very long period, it undergoes mummification. It must be borne in mind that the body of a macerated fœtus,

Marginal note: Intra-uterine maceration.

when exposed to the air outside the body, will undergo the ordinary putrefactive changes with great rapidity, and soon lose the distinctive characters of the intra-uterine maceration.

If the body of a newly-born child is found dead, but still covered with the vernix caseosa, there would be a presumption that the child might have been still-born. The presence of suggillations, or bruise-like discolourations, have no value as a means of diagnosis. When putrefaction is far advanced, it will be impossible to decide whether a child has been still-born or not, unless the osseous nucleus at the lower end of the femur, which is but little affected by putrefaction, should afford positive evidence of life after birth.

It must be admitted that there are cases in which, legally, a child might be adjudged to have lived after its birth, though no medical proof of that fact would be obtainable, by an examination after death. This remark is especially applicable to such cases as that quoted on page 357, where a mere tremulous motion of the lips was accepted as sufficient evidence of live birth. It is clear, then, that in the case of a child which had reached the full term, it would be impossible to assert positively that it had never lived, simply because there was no evidence of its having breathed. A post-mortem examination cannot really settle the question as to whether a child has lived for a few minutes or not, but, when the child has survived a few hours or days, certain changes take place which leave indelible evidence behind them. It should be noted that marks of violence may be present without having been caused during life, and without necessarily being the cause of death. Violence applied to a child with the intention of putting an end to its existence is, as a rule, unnecessarily excessive.

Premising that the presence of either warmth

or rigor mortis in the body of a newly-born child when found, does not point to live birth of necessity, the condition of the organs will be examined in the following order: the skin, the umbilical cord, the brain, the internal ear, the alimentary canal, the urinary apparatus, the circulatory system, and the respiratory system. After that the causes of death, both natural and violent, in the new-born infant will be considered.

Signs of life after birth.
At birth, the *skin* is of a dark red colour, but almost at once the colour commences to grow lighter, remaining so for about a day, and then getting slightly darker again for a couple of days, after which it gradually (by the end of the first week) assumes the reddish-white colour which it retains throughout life. Between the first and the fourth days, the epidermis gradually scales off, either as a fine dust or in layers; this scaling commences on the abdomen, and spreads thence in all directions; it is not usually complete for several weeks. It must not be confused with the peeling which has already been described, as taking place during intra-uterine maceration.

In examining the *umbilical cord*, the first thing is to notice whether it has been ruptured or cut, and, if the latter, whether evenly or not. If the cord has obviously been ruptured, the fact would support the theory of self-delivery; if the cord had been cut and tied, there would be a presumption, if not of assistance at the labour, at any rate that the woman had been quite prepared for it, which might be in contradiction of her own statements. If pulsation were recognised in the cord after the birth of the entire body, this would be conclusive of live birth. After the cord has been tied and divided, it undergoes certain changes—supposing, that is, that the child lives—resulting in a process of desiccation, which leads to its finally dropping off. Within a few hours after the application of the ligature, it loses its smooth, shiny appearance, and becomes dry and flaccid. Desiccation, or mummification, commences in about twenty-four hours at the ligatured end, and gradually spreads along the cord towards the umbilicus; it becomes shrivelled, twisted, and about the third day is of a yellowish colour, and semi-translucent. About the third day a zone of capillary congestion appears, indicating the future line of demarcation; and, usually between the fourth and sixth days, the dried portion falls off at this spot, which is situated close to the umbilical end of the cord. After this, cicatrisation of the umbilicus takes place, and should be completed between the tenth and fourteenth days after birth. These changes can none of them take place in the dead body, and the mummification and desiccation of the cord must not be confounded with the ordinary putrefactive changes.

The state of the *brain* does not afford very valuable evidence, and unless the body were examined soon after death, it would have no value at all. In the child born before the full term there is no differentiation between the grey and the white matter; the whole brain appears homogeneous. At the full term there will be found, at the site of the fissure of Rolando, some violet fibres, and a fortnight later, commencing differentiation into grey and white matter is shown, in that the formation of the white strands composing the internal capsule can be recognised. These changes were made out by the careful study of a large number of brains, by the late M. Parrot.

A few years ago attention was called to the state of the *internal ear*, as affording good evidence of the fact of respiration. Before respiration, the internal ear is filled with a gelatinous pulpy material, formed of mucous membrane in an embryonic state.

After respiration, this undergoes rapid absorption, and is replaced by air. Tidy, however, denies that any practical value attaches to this sign, since, in his experience, the replacement of the gelatinous material by air has varied within such wide limits as a few hours and five weeks.

The state of the *alimentary canal* may afford very valuable evidence, on the question of live birth. In a still-born child the stomach may contain some of the liquor amnii, and also a peculiar substance, probably in great measure derived from the salivary secretion, differing, both in characters and quantity, at different periods of fœtal life. Besides the above, meconium may also be found in the fœtal stomach. Meconium (so called on account of its resemblance in colour to the juice of the poppy) is an olive-green substance of thick consistence, and without odour. It commences to be formed in the intestines of the fœtus, about the sixth month of intra-uterine life, and is expelled during the first few days after birth; when discovered in the stomach it is probable that it has been swallowed after being passed from the bowel before birth, and affords no indication that the child has been born alive. On the other hand, the state of the alimentary canal may afford conclusive evidence of live birth. Thus, as regards the presence or absence of meconium, Tidy points out that its entire absence from the intestines affords very strong evidence that the child was born alive, and survived its birth some time. The presence of a large quantity of meconium, in the large and small intestines, would afford presumptive evidence that the child had not lived. The presence of air in the stomach, especially if intimately mixed with its contents, is a proof that respiration has taken place. Breslau of Prague draws the following conclusions on this subject:—1. In children born

INFANTICIDE.

dead, independently of whether the death occurred before or during birth, no air is to be found in the stomach or intestines. If, therefore, the stomach and intestines of a still-born child be removed (after they have been carefully tied and secured) and placed in water, they will sink. 2. The presence of air in the stomach depends upon respiration (the air being swallowed during inspiration), and is independent of the taking of food. Hence the air probably reaches the stomach with the first respiration, and, as breathing proceeds, finds its way by degrees into the intestines. 3. After respiration, the stomach and intestines, when placed in water, float. 4. The more completely the intestines are inflated, and the lower in the bowels that air is found, the longer, in all probability, the child has lived, and the more certain is the evidence of live birth.

The discovery of starch, sugar, or milk, in the stomach would be almost conclusive of live birth, as it would be in the highest degree improbable that the child could be fed before it was entirely born. Blood corpuscles may be found in the stomach, the blood having been possibly swallowed during the delivery, and before the entire birth of the child. The liver loses weight, at the time of birth, with the establishment of respiration, as some of the blood, that before went to it, then goes to the lungs; but its weight is of no value as a test of live birth, as the difference is only relative, and a comparison with the weight before birth in any particular case is impossible.

In regard to the *urinary organs*, no reliance can be placed on the state of the bladder; it may be found empty in a still-born child, and full in one that has survived its birth. Uric acid crystals, in the form of golden yellow streaks along the papillæ, in the kidneys, are generally to be seen in infants that have lived a few days, and they have been regarded as affording conclusive evidence of life after birth. They are supposed to be formed by the rapid oxidation of the tissues, which takes place after the establishment of respiration, the urine, just at first, not being secreted in sufficient quantity to remove them. Uric acid infarction, as this is called, may, according to Tidy, be occasionally found in the bodies of still-born children, and is not universally present in the bodies of those that have lived a short time, so that too much reliance must not be placed on this sign, which nevertheless has a certain value.

From the moment of birth, changes at once commence in the *circulatory system*. These changes relate to the umbilical vessels, the ductus arteriosus, the ductus venosus, and the foramen ovale. The umbilical vessels are gradually closing whilst the cord is undergoing desiccation, so that by the time this is accomplished, and the cord drops off, the vessels in all ordinary cases are completely obliterated, and hence there is no hæmorrhage from them. This closure takes place more rapidly in the arteries than in the veins, and in both appears to be due to a concentric thickening, so that whilst there may be no outward change in the size of the vessel, the lumen is found to be much reduced until it is ultimately occluded.

The ductus arteriosus, during fœtal life, conveys blood from the right ventricle, by means of the pulmonary artery, to the aorta, but after the establishment of respiration the blood is conveyed in much larger quantities to the lungs, and the need for the ductus arteriosus ceases; it therefore begins to become obliterated. According to Prof. Bernt of Vienna the closure generally commences at the aortic end of the duct, which appears contracted, so that the whole tube has a conical shape;

a little later the cardiac end also contracts, and the tube again becomes cylindrical. He found that the duct was not closed until after one or more weeks, but cases have been recorded, in which the duct was found impervious within a few minutes of birth. Its patency, even at an advanced age, is by no means uncommon, and is found in connection with malformation of the heart, where the orifice of the pulmonary artery is occluded, the blood then entering the latter from the aorta.

The ductus venosus runs from the umbilical vein towards the liver, to open directly into the inferior vena cava. After birth it collapses, and about the second or third day it begins to be obliterated, but the exact period at which this process is completed is somewhat uncertain. There does not appear to be any case on record, in which it was found closed at birth.

The foramen ovale is the communication which exists during fœtal life, between the right and left auricles of the heart. After birth, this communication is no longer required, and the foramen gradually closes up. Its closure is usually complete about the tenth day. In a few cases it has been found closed at the time of birth, and its permanent patency is one of the common forms of congenital malformation of the heart. These changes in the circulation have been known as the docimasia circulationis (from δοκιμάζω, I examine); it will be seen that the several obliterations are generally complete by the tenth day. They take place in the following order:—the umbilical arteries, the ductus venosus, the ductus arteriosus, and the foramen ovale.

The state of the *respiratory apparatus* requires a more detailed consideration, as, although it has been repeatedly mentioned that proof of breathing is not essential to the establishment of live birth, yet the degree of respiration may afford conclusive evidence on the point. During fœtal life the larynx is narrow and filled with mucus, and the vocal cords lie in close apposition; after respiration the glottis becomes opened up. The trachea before respiration is flattened.

State of the lungs before respiration. If the thorax be opened in the usual manner, the following will be the appearances: If respiration has not taken place, the diaphragm will be arched at a high level, perhaps the fourth or fifth rib, the thymus and pericardium will come prominently into view, occupying the whole of the middle of the thorax, the heart inclining slightly to the left; the lungs will not be visible at all. On further examination they will be found lying at the back of the thorax, on each side of and close to the vertebral column; they are longer than they are broad, of a dark maroon or brownish-red colour, and feel firm. They do not crepitate when squeezed, but may be torn by compression. *After complete respiration.* When complete respiration has taken place, the change on opening the thorax is most striking: the thymus occupies a small area at the upper part of the mediastinum; the pericardium is in great part covered by the lungs, which are of a light red or mottled hue, patches of darker colour being often found intermingled with the light-coloured parts; they completely fill the thorax, and, when pressed, they crepitate under the finger. When such appearances as these are found, there can be no doubt that the child has been born alive, and that respiration has taken place very fully; but between these two extremes all degrees of imperfect respiration may be met with. *After partial respiration.* When only very imperfect respiration has taken place, a few slightly-raised bright red patches will be seen here and there on the surface of the lungs; these are the inflated air vesicles. When none can be seen with the naked eye, the surface of the lung might be examined with the aid of a lens. These appearances are the same after artificial respiration as after natural breathing.

There are several tests, all of more or less value in determining the degree to which the lung has been expanded by air, and the importance to be attached to that degree. They are:—

1. *The Static Test.* The absolute weight of the lungs after respiration is greater than before. This is due to the increased amount of blood they receive, consequent upon their increased capacity. It has been pointed out that the lung which has been artificially inflated weighs actually less than it did whilst in the fœtal state, as some blood would have been expelled during the process. 2. *Ploucquet's Test.* This is founded upon a comparison between the absolute weight of the body and of the lungs. M. Ploucquet stated, as the result of his investigations, that the weight of the lungs to the body in children which had not breathed was, as 1 to 70, and in those which had, as 2 to 70. The limits of error, however, in this calculation are so wide that the test has no practical value. 3. *The Hydrostatic Test.* The specific gravity of the lungs is diminished by breathing, although the actual weight is increased. This is owing to the far greater bulk of air they contain. The aërated lung therefore floats on water, and upon this fact the most reliable test has been founded. When complete respiration has taken place, the lungs will be so buoyant that they will float even while still attached to the other thoracic viscera. If they do not float, the lungs should be separated from everything else and each other, and tested by themselves; if they still sink, each should be divided into small portions, and these thrown in one by one; if each of these sinks, it is certain that respiration has not taken place. If the lungs

Tests.

when whole or divided float, the pieces should severally be subjected to firm compression; and if, after as much air as possible has been driven out in this way, they still float, then the fact of respiration is established beyond doubt. It must not be forgotten, however, that the absence of proof of respiration is no proof of still-birth, and the proof of respiration is not conclusive that the child was born alive.

Objections to the hydrostatic test. Other objections, too, have been raised to this test. The lungs may sink from disease—inflammation of the lung will certainly cause it to sink; or there may be a congenital non-expansion of the lung, a condition known as atelectasis, in which, from some as yet unexplained cause, a greater or less area of the lung remains in its fœtal state, though on examination it can easily be artificially inflated, and no mechanical cause for the failure to expand can be detected. A case has been recorded, in which the heart could be felt to pulsate for sixteen minutes after birth, but there was no respiration, and at the autopsy the lungs showed no signs of expansion, and did not float. Again, the lungs may float either from putrefaction or from artificial inflation. The development of gas in the lungs, as a result of putrefaction, is of course possible, but it should be accompanied by signs of equally advanced putrefaction in the other viscera, and, when this is the case, it is generally impossible to determine whether the air found in the lungs is due wholly to this cause, or whether some of it may not have been derived from respiration. As the lungs cannot be inflated fully with air whilst in the body of a new-born child, it is only in case of imperfect respiration that artificial inflation could be called in to account for the presence of the air found. Air introduced in this way can, as a rule, be entirely expelled by compression of the lung.

It is doubtful, however, whether much air could be introduced into the lungs of a still-born child, except by one experienced in the necessary manipulation. The only cases in which this point has to be determined are those where it is a question, whether a mother has killed her child, or whether it was still-born, and it seems improbable that she should take steps to make it appear that the child had been born alive, when it would be in her own interest to prove the opposite. Tidy sums up the conclusions in the following tabular manner:—

Lungs which have not breathed.	Lungs which have breathed.
1. Dark in colour (black-blue, maroon, or purple), resembling liver.	1. Light in colour (rose-pink, pale pink, light red, or crimson).
2. Air-vesicles not visible to the naked eye.	2. Air-vesicles distinctly visible to the naked eye, or to a lens of low power (say a two-inch, or even a common reading glass).
3. Do not crepitate or crackle when squeezed or cut.	3. Crepitate or crackle freely.
4. Contain but little blood, therefore little escapes on section.	4. Contain a good deal of blood, which escapes freely on section.
5. The blood present is not frothy, unless there be putrefaction.	5. This blood is freely mixed with air, and therefore appears frothy.
6. Sink in water, unless putrid, and often even then.	6. Float' in water, or at all events, the parts which have been expanded, or have breathed, float. If fully expanded, they will even buoy up the heart.
7. Bubbles of gas arising from putrefaction may be squeezed out, and as they escape are usually noted to be of large size.	7. The air cannot be squeeced out.

Other signs, which have been mentioned in connection with the respiratory system, concern the amount of blood found in the pulmonary vessels, and the relative proportion of fat in the lungs, but no inferences of value can be drawn from either of these.

In the event of the discovery of the body of a

Estimation of the age of an infant.

new-born infant, which has evidently survived its birth, it may be of great importance, as an aid to the detection of the mother, to determine how long the child lived. On this point the following conclusions are given by Tidy:—If the child survived from *a few minutes* to *some hours*: the stomach contains a frothy fluid, and clots will be found in the vessels of the umbilical cord. After the *first* day: contraction and thickening of the coats of the umbilical arteries near the umbilicus. After the *second* day: contraction throughout the greater part of the umbilical arteries; the epidermis begins to exfoliate. After the *third* day: contraction of the umbilical arteries to their termination in the iliacs; slight contraction in the umbilical veins; desiccation of the cord, the formation of an inflamed ring, and a slight purulent discharge at the point of ultimate separation. After the *fourth* day: the cord separates. After the *fifth* day: contraction of the umbilical veins complete. After the *seventh* day: the ductus arteriosus contracted to the size of a crow quill. From the *eighth* to the *tenth* days: the fœtal openings (ductus arteriosus, ductus venosus, foramen ovale) become obliterated. From the *tenth* to the *twelfth* days: the osseous centre of the femoral epiphysis measures more than from five to six millimetres in diameter; cicatrisation of the umbilicus. (If the umbilicus be healed, it indicates life for about twenty-one days.)

Causes of death.

The causes of death in the new-born infant alone remain to be considered. These are usually divided into those which are natural, and those resulting from violence, whether accidental or intentional. Amongst the former, the first place must be given to cases where death takes place during delivery. This may be due to simple prolongation of the labour, but more often is the result of some abnormal position of the cord in relation to the child's head, so that it is compressed, and the fœtal circulation thus completely interrupted; or the cord may be so firmly twisted, or knotted, during the movements of the fœtus in the uterus, that death may result.

Amongst the natural causes of death after birth may be mentioned, in the first place, malformations. It is quite conceivable that the child might have such a malformation as to render it impossible, that it should maintain a separate existence at all, though fœtal life might have been maintained without any hindrance; such cases, however, are not common. An immature child, in point of time, will be proportionally immature as regards development, and therefore less able to withstand the demands made upon its vital powers. Premature children are very liable to die from mere feebleness; nor is death from simple debility confined to them, as those born at the full time are sometimes so feeble as to require the most sedulous care, and it can readily be understood that such children might die, if the mothers were without aid at the time of their confinements.
_{Natural causes.}

Amongst diseased conditions which cause the death of the new-born infant, those which affect the respiratory tract are the most frequent and the most fatal. Inflammations of the lungs, bronchi, or pleuræ are the chief of these, especially when they are accompanied by extensive collapse of portions of the lung; spasm of the glottis is often given as one of the causes. Epileptiform convulsions, or meningeal hæmorrhage, may also cause speedy death after birth.
_{Death from disease.}

Hæmorrhage from the cord may occur from its being ruptured during delivery, and not tied subsequently. This, however, seldom causes death, as it is well known that a vessel which has been torn has no great tendency to bleed. Occasionally a fatal oozing has taken place after the cord was tied, owing to the ligature having been inefficiently applied; this, too, it must be admitted, rarely happens. Secondary hæmorrhage at the time of the separation of the cord has been known to be fatal; in such cases there was probably a tendency to the hæmorrhagic diathesis.

Taylor arrives at the following conclusions on this subject: 1. A large number of illegitimate children, especially when immature, are born dead from natural causes. 2. A child may die from exhaustion, as the result of a protracted labour. 3. If a child is prematurely born, or if it is small, and weak, even at the natural period, it may die from mere debility, or want of power in the constitution, either to commence, or to continue, the act of respiration. 4. A child may die from loss of blood owing to the accidental rupture of the cord during delivery; it may even die from this cause after it has breathed. 5. Fatal bleeding is more likely to occur when the cord has been cut close to the abdomen than when it has been lacerated, or cut, at a distance from the navel. 6. A division of the cord, whether by rupture or incision, without ligature, is by no means necessarily fatal to a healthy mature child. 7. A child may die from accidental compression of the cord during delivery, the circulation between the mother and child being thereby arrested, before or after breathing.

Death from neglect. The causes of death which cannot be accounted as natural are several. Thus, neglect on the part of the mother will infallibly lead to the death of the child; by neglect would be understood the omission to duly provide, not only food, but clothing and warmth. Exposure to cold is one of the causes of death most difficult of proof; the withholding a due supply of nourishment might be inferred from the empty state of the alimentary canal after death, and, if it were urged that this was the result of vomiting and diarrhœa, a careful microscopical examination of the contents of the stomach and œsophagus would show whether this theory could be supported, by the discovery of milk corpuscles or starch granules; the absence of these would point to starvation.

Drowning is one of the causes of death in the newly-born. If respiration had taken place, the signs would not differ from those already described as found in the adult. If the child had not breathed, before being thrown into the water, the fact of drowning would be very difficult to prove. Not infrequently the child is put into a privy, in the soil of which it is speedily drowned or suffocated; in such a case it might be urged that the woman was accidentally delivered whilst straining at stool; this would be corroborated if the cord showed signs of having been torn, and not cut. *Drowning.*

Strangulation is, perhaps, the method most often adopted in cases of child-murder. This may occur from a natural cause, owing to the funis being round the neck at the time of delivery, as already described. When this happens, however, the mark will not be very well characterised, nor should there be much damage done to the deeper structures; further, on examination, the lungs should only show the very slightest, if any, traces of respiration, as it is obvious that the strangulation would take place almost immediately after birth. There should, therefore, be very little difficulty in excluding accidental strangulation due to this cause. The mark of a ligature would present the same characters, as those which have been already mentioned in reference to adults. Ecchymoses might also be produced in the skin, and, in the case of manual violence, imprints of the fingers might be easily recognised. It must be borne in mind, however, that these marks may be produced a short *Strangulation.*

time after death, and that they may also be produced before complete and entire birth, so that it is hardly ever possible for the medical witness to say, positively, that the child was living at the time the violence was inflicted.

Suffocation. Suffocation is at once the easiest method of putting an end to a child's existence, and the most difficult of recognition. It may be accidental; portions of food or vomit may get inhaled into the larynx and obstruct it, or the child may be put carelessly into bed, and be suffocated by the pillow or bedclothes. The bare possibility of accidental suffocation in a privy has just been alluded to. Feathers, wool, dust, &c., have, at one time or another, been stuffed into an infant's mouth with this object in view. Suffocation may be effected simply by the hand, held in front of the mouth for a minute or so. The proof of death by suffocation is very difficult, if nothing can be discovered in the mouth or larynx. A good deal of stress has been laid upon the existence of subpleural ecchymoses.

Wounds. Wounds on the body of the new-born infant present the same characters as those in the adult. Incised or punctured wounds, and others more rarely, may be attributed, sometimes truly, to the efforts of the woman at self-delivery. The blood-stained œdematous swelling on the scalp, known as the *caput succedaneum*, the result of prolonged labour, must not be mistaken for the consequences of violence.

Fracture of the skull. A fracture of the skull may be the result of violence applied to the fœtus during intra-uterine life; or it may be found, especially in the parietal bones, as the result of the forcible pressure of the head against the pelvis during labour; or the parietal bones may be fissured by the force of the uterine contractions compressing the head; or the application of forceps, or other instruments, to aid in delivery may cause a fracture of the cranial bones. Again, the skull may be fractured by the delivery being sudden, and the child falling on to the floor. The elasticity of the bones of the skull is, however, so great that it is not often that, under these circumstances, a fracture would take place. The cord should be examined, as, if it had been cut, it would be evident that the statement, that delivery had been sudden, was untrue, and that the fall, if it took place at all, did not take place during delivery. One general remark, which has already been made about murders, may be here repeated, viz. that the violence is usually greatly in excess of that necessary to destroy life.

A wound or fracture produced just after death, could not be distinguished from one caused during life, but in the former case the cause of death should be obvious, and there should be a satisfactory explanation at hand of how a recently dead body came to be wounded in that particular way. In a few very rare instances new-born children have been put to death by the administration of poison.

INDEX.

ABD

ABDOMINAL enlargement in pregnancy, 307
Abdominal injuries, 141
Abortion, causes of, 327
—— criminal, 324
—— signs of, 332
—— statute relating to, 324
Accidents, insurance against, 162
Acetic acid, poisoning by, 212
Acids, the mineral, 190
Aconite, poisoning by, 297
Aconitine, poisoning by, 298
Adipocere, 32
Æthusa cynapium, poisoning by, 283
Age of a skeleton, determination of, 42
Age, legal relationships of, 359
Air, influence of, on putrefaction, 29
Alcohol, acute poisoning by, 265
Alcoholism, diagnosis of, from opium poisoning, 256
Alimentary canal, changes in the, after birth, 368
Alkalies, the, 201
Alkaloids, cadaveric, test for, 301
Alum, poisoning by, 208
American crowbar case, the, 134
Ammonia, poisoning by, 203
Anæsthetics, death during the administration of, 129
Aniline, poisoning by, 263
Animal irritants, poisoning by, 248
Antimony, poisoning by, 228
Arsenic, poisoning by, 220
Arsenical wall-papers, 226
Asphyxia, 69
Atelectasis, 374

CHE

BARIUM salts, poisoning by, 208
Barrel, examination of the contents of, 146
Belladonna, poisoning by, 258
Bichromate of potash, poisoning by, 244
Birth, live, 356
Bisulphide of carbon, poisoning by, 272
Bladder, rupture of, 142
Blisters as a result of burns, 151
Blood-stains, 116
Brain, changes in the, after birth, 367
Brain, injuries to the, 133
Bravo, Mr., case of, 229
Breast, changes in the, in pregnancy, 306
Bromide of potassium, poisoning by, 219
Bromine, poisoning by, 219
Burns and scalds, 149

CADAVERIC ecchymoses, 14
—— spasm, 19
Camphor, poisoning by, 262
Cantharides, poisoning by, 248
Carbolic acid, poisoning by, 273
Carboluria, 274
Carbon bisulphide, poisoning by, 272
Carbonic acid, poisoning by, 302
Carbonic oxide, poisoning by, 303
Castration, 347
Catamenia, cessation of, in pregnancy, 305
Certificates of lunacy, 173
Chest, penetrating wounds of the, 139

CHL

Chloral hydrate, poisoning by, 270
Chloride of antimony, poisoning by, 230
Chloroform as an anæsthetic, 268
Chronic poisoning by arsenic, 227
—— —— by copper, 233
—— —— by lead, 240
—— —— by phosphorus, 217
Cicatrices, 146
Cicuta virosa, poisoning by, 283
Circulation, cessation of, 9
—— changes in the, at birth, 370
Coal gas, poisoning by, 303
Cocculus indicus, poisoning by, 288
Colchicum, poisoning by, 247
Cold, death from, 61
Cole, case of, 180
Colour vision, defective, 165
Combustion, spontaneous, 153
Concealment of pregnancy, 312
Conium, poisoning by, 281
Contracts, insanity in relation to, 168
Contused wounds, 107
Cooling of the body, 11
Copper, poisoning by, 233
—— utensils, poisoning from the use of, 234
Corpus luteum, its value, 320
Corrosive sublimate, poisoning by, 235
Cotton fibres, characters of, 115
Crime, insanity in relation to, 178
Croton oil, poisoning by, 247
Crying as a test of live birth, 358
Curare, poisoning by, 284

DEATH, signs of, 8
—— sudden, causes of, 52
Delirium tremens, legal relationships of, 182
Delivery at a remote period, signs of, 316
Delivery, signs of past, 322
—— signs of recent, 314
Diabetic coma, diagnosis of, from opium poisoning, 256
Digitalis, poisoning by, 295
Disease in relation to poisons, 186
Dislocations, 143
Disputed paternity, 342
Douglas case, the, 342
Drowning, 70
—— in the new-born infant, 379
Ductus arteriosus, date of closure of, 370

HEA

Duncan, Matthews, on sterility, 351
Dwight case, the, 161
Dying declarations, 4

EAR, internal changes in the, after birth, 367
Ecbolics, 329
Ecchymoses, cadaveric, 14
—— made during life, 106
Embryo, development of the, 322
Emmenagogues, 329
Entrance, wounds of, characters, 145
Epileptic mania, 181
Ergot as an abortive, 331
—— poisoning by, 248
Ether as an anæsthetic, 267
Evidence given by the insane, 172
Exit, wounds of, characters, 145

FACE, injuries to the, 135
Feigned death, in relation to life insurance, 164
Feigned insanity, 181
—— pregnancy, 311
Flaccidity as a sign of death, 17
Fœtal heart, sounds of the, 310
Foods, poisonous, 249
Foramen ovale, closure of the, 371
Fractures, spontaneous, 143
Fungi, poisonous, 263

GARDNER peerage case, 339
Garfield, President, case of, 127
Gases, poisonous, 302
Genital organs, wounds of, 142
Gestation, short duration of, 336
Good v. Whittle, 175
Gouldstone, case of, 179
Guiteau, case of, 180
Gunshot wounds, 144

HABIT in relation to poisons, 186
Hæmorrhage as a cause of death, 122
Hæmorrhage, cerebral, diagnosis of, from opium poisoning, 256
Hæmorrhage from the umbilical cord, 377
Hair adhering to a weapon, 115
Halpin's (Dr.) case, 336
Hanging, 88
Head, injuries to the, 132
Heart, wounds of the, 139
Heat, death from, 63

INDEX

HEI

Height, estimation of the, 48
Hellebore, poisoning by, 247
Hermaphroditism, 343
Hopley, case of, 12
Hydrochloric acid, poisoning by, 198
Hydrostatic test, 873
Hyoscyamus, poisoning by, 260

IDENTITY, 37
Idiosyncrasy in relation to poisons 185
Impotence and sterility in the female, 349
Impotence and sterility in the male, 346
Incised wounds, 107
Indirect causes of death from wounds, 126
Infanticide, 361
Inheritance, 358
Insanity, definition of, 167
Insensibility, 10
Insurance, evils of, 163
—— life, 154
Intemperance in relation to life insurance, 157
Interdiction, 175
Intoxication, legal relationships of, 182
Iodide of potassium, poisoning by, 218
Iodine, poisoning by, 218
Iron, poisoning by, 243
Irritability, muscular, extinction of, 17
Irritant poisons, 190

KIESTEIN, 308

LACERATED wounds, 107
Lamson, the trial of, 299
Lane, Harriet, murder of, 39
Lead colic, 240
—— palsy, 241
—— poisoning by, 239
Legge v. Edmunds, 348
Life insurance, 154
Lightning, death from, 99
Linen fibres, characters of, 115
Lobelia, poisoning by, 289
Lunacy certificates, 173
Lungs, condition of, before respiration, 372

POI

MACERATION, intra-uterine, 364
Madness and crime, 178
Malapraxis, charge of, 129
Maturity in the new-born infant, 337
Maturity in the new-born infant, signs of, 361
Mercurial tremor, 237
Mercury, poisoning by, 235
Microscopical examination of blood, 119
Mind, a disposing, 169
Moisture, influence of, on putrefaction, 28
Monsters, 341
Morning sickness in pregnancy, 306
Mummification, 36
Mutilated remains, 37
Myosin, formation of, 19

NECK, injuries to the, 136
—— the mark on the, in hanging, 90
Neurotic poisons, 252
Nicotine, poisoning by, 288
Nitre, poisoning by, 205
Nitric acid, poisoning by, 196
Nitro-benzine, poisoning by, 263
Nitro-glycerine, poisoning by, 295

ŒNANTHE crocata, poisoning by, 282
Old age, wills made in, 170
Opium eating, 258
—— poisoning by, 252
Osseous nucleus at lower end of femur, 363
Outrepont's (Dr.) case, 336
Oxalic acid, poisoning by, 209

PALMER, William, case of, 278
Pelvis, female, characters of the, 51
Phosphorus, poisoning by, 214
Physostigma faba, poisoning by, 285
Ploucquet's test, 373
Plural births, 342
Poisoning, general evidence of, 187
Poisons, causes modifying their action, 185
Poisons, medical definition of, 184
—— mode of action of, 184
—— statutes relating to, 183

C C

POS

Post-mortem bleeding, 25
—— delivery, 318
—— digestion of the stomach, 26
Potash, bichromate of, poisoning by, 244
Potash, caustic, poisoning by, 201
—— chlorate of, poisoning by, 206
—— sulphate of, poisoning by, 207
Potassium, bromide of, poisoning by, 219
Potassium, iodide of, poisoning by, 219
Pregnancy, duration of, 334
—— signs of, 305
Presumption of death, 53
—— of survivorship, 56
Pretended death, 55
Prolonged gestation, 338
Prussic acid, poisoning by, 290
Ptomaines, 250
Puerperal mania, 181
Punctured wounds, 108
Putrefaction, 23
—— delayed by arsenic, 223
—— in the earth, 30
—— in water, 33

QUICKENING, 309

RAILWAY spine, the, 137
Rape, 352
Respiration, cessation of, 9
Rigor mortis, 18

SALIVATION, mercurial, 237
Savin as an abortive, 330
—— poisoning by, 246
Scars from disease, 148
Scott, Gilbert, case, 176
Self-inflicted wounds, 130
Semple, Weldon v., 174
Sex, determination of, in the skeleton, 50
Shock, death of wounded persons from, 123
Siamese twins, 341
Silk, microscopical appearances of, 115
Silver, nitrate of, poisoning by, 243
Skin, changes in the, after birth, 366
Soda, caustic, poisoning by, 201
Spectrum analysis of blood, 119
Stains resembling blood, 117
Starvation, 64

UXB

State control of lunatics, 173
Staunton case, the, 67
Sterility and impotence in the female, 349
Sterility and impotence in the male, 346
Stramonium, poisoning by, 261
Strangulation, 93
—— in the new-born infant, 379
Strychnia, poisoning by, 276
Suffocation, 96
—— in new-born infants, 380
Suicidal wounds, characters of, 109
Suicide in relation to life insurance, 160
Sulphuretted hydrogen, poisoning by, 304
Sulphuric acid, poisoning by, 190
Superfœtation, 339
Supposititious children, 342
Syphilis as a cause of abortion, 327

TARTAR emetic, 228
Tartaric acid, poisoning by, 212
Tattooing, 149
Teeth in reference to identity, 44
Temperature, influence of, on putrefaction, 28
Temperature, loss of, 10
Tenancy by courtesy, 359
Tetanus contrasted with strychnia poisoning, 280
Tichborne case, 147
Tobacco, poisoning by, 287
Turpentine, poisoning by, 295

UMBILICAL cord, changes in the, after birth, 366
Unconscious delivery, 317
—— impregnation, 312
Underwood v. Wing, 58
Unnatural offences, 355
Unsound lives in life assurance, 155
Uræmic coma, diagnosis of, from opium poisoning, 256
Urinary organs, state of, at birth, 369
Uterus, changes in the, in pregnancy, 308
Uterus, characters of the virgin, after death, 311
Uterus, state of, after delivery, 318
Uxbridge murder, the, 111

INDEX.

VEG

VEGETABLE irritants, 246
Vertebræ, injuries to the, 136
Viability, 336
Violence as a means of procuring abortion, 328
Vision, 165

WALSH, Caroline, case of, 46
Water, action of, on lead, 242
Weldon v. Semple, 174
Welsh fasting girl, the, 66
Wells, Sir S., on sterility, 350

ZIN

Whittle, Good v., 175
Wills, insanity in relation to, 169
Woollen fibres, characters of, 115
Wounds, classification of, 103
—— definition of, 102

YELLOW atrophy of the liver, acute, 215
Yew, poisoning by, 284

ZINC salts, poisoning by, 242

[CATALOGUE C]

LONDON, *November*, 1884.

J. & A. CHURCHILL'S

MEDICAL CLASS BOOKS.

ANATOMY.

BRAUNE.—**An Atlas of Topographical Anatomy**, after Plane Sections of Frozen Bodies. By WILHELM BRAUNE, Professor of Anatomy in the University of Leipzig. Translated by EDWARD BELLAMY, F.R.C.S., and Member of the Board of Examiners; Surgeon to Charing Cross Hospital, and Lecturer on Anatomy in its School. With 34 Photo-lithographic Plates and 46 Woodcuts. Large Imp. 8vo, 40s.

FLOWER.—**Diagrams of the Nerves of the Human Body**, exhibiting their Origin, Divisions, and Connexions, with their Distribution to the various Regions of the Cutaneous Surface, and to all the Muscles. By WILLIAM H. FLOWER, F.R.C.S., F.R.S. Third Edition, containing 6 Plates. Royal 4to, 12s.

GODLEE.—**An Atlas of Human Anatomy:** illustrating most of the ordinary Dissections and many not usually practised by the Student. By RICKMAN J. GODLEE, M.S., F.R.C.S., Assistant-Surgeon to University College Hospital, and Senior Demonstrator of Anatomy in University College. With 48 Imp. 4to Coloured Plates, containing 112 Figures, and a Volume of Explanatory Text, with many Engravings. 8vo, £4 14s. 6d.

HEATH.—**Practical Anatomy: a Manual of Dissections.** By CHRISTOPHER HEATH, F.R.C.S., Holme Professor of Clinical Surgery in University College and Surgeon to the Hospital. Fifth Edition. With 24 Coloured Plates and 269 Engravings. Crown 8vo, 15s.

11, *NEW BURLINGTON STREET.*

J. & A. Churchill's Medical Class Books.

ANATOMY—*continued.*

HOLDEN.—A Manual of the Dissection of the
Human Body. By LUTHER HOLDEN, F.R.C.S., Consulting-Surgeon to
St. Bartholomew's Hospital. Fifth Edition, by JOHN LANGTON,
F.R.C.S., Surgeon to, and Lecturer on Anatomy at, St. Bartholomew's
Hospital. With 208 Engravings. 8vo, 20s.

By the same Author.

Human Osteology : comprising a Description of the Bones, with Delineations of the Attachments of the
Muscles, the General and Microscopical Structure of Bone
and its Development. Sixth Edition, revised by the Author and
JAMES SHUTER, F.R.C.S., late Assistant-Surgeon to St. Bartholomew's Hospital. With 61 Lithographic Plates and 89 Engravings.
Royal 8vo, 16s.

ALSO,

Landmarks, Medical and Surgical. Third
Edition. 8vo, 3s. 6d.

MORRIS.—The Anatomy of the Joints of Man.
By HENRY MORRIS, M.A., F.R.C.S., Surgeon to, and Lecturer on Anatomy and Practical Surgery at, the Middlesex Hospital. With 44
Plates (19 Coloured) and Engravings. 8vo, 16s.

The Anatomical Remembrancer; or, Complete Pocket Anatomist. Eighth Edition. 32mo, 3s. 6d.

WAGSTAFFE.—The Student's Guide to Human
Osteology. By WM. WARWICK WAGSTAFFE, F.R.C.S., late Assistant-Surgeon to, and Lecturer on Anatomy at, St. Thomas's Hospital.
With 23 Plates and 66 Engravings. Fcap. 8vo, 10s. 6d.

WILSON — BUCHANAN — CLARK. — Wilson's
Anatomist's Vade-Mecum : a System of Human Anatomy. Tenth
Edition, by GEORGE BUCHANAN, Professor of Clinical Surgery in the
University of Glasgow, and HENRY E. CLARK, M.R.C.S., Lecturer on
Anatomy in the Glasgow Royal Infirmary School of Medicine. With
450 Engravings, including 26 Coloured Plates. Crown 8vo, 18s.

11, *NEW BURLINGTON STREET.*

J. & A. Churchill's Medical Class Books.

BOTANY.

BENTLEY.—A Manual of Botany. By Robert
BENTLEY, F.L.S., M.R.C.S., Professor of Botany in King's College and to the Pharmaceutical Society. With 1185 Engravings. Fourth Edition. Crown 8vo, 15s.

By the same Author.

The Student's Guide to Structural,
Morphological, and Physiological Botany. With 660 Engravings. Fcap. 8vo, 7s. 6d.

ALSO,

The Student's Guide to Systematic
Botany, including the Classification of Plants and Descriptive Botany. With 357 Engravings. Fcap. 8vo, 3s. 6d.

BENTLEY AND TRIMEN.—Medicinal Plants:
being descriptions, with original Figures, of the Principal Plants employed in Medicine, and an account of their Properties and Uses. By ROBERT BENTLEY, F.L.S., and HENRY TRIMEN, M.B., F.L.S. In 4 Vols., large 8vo, with 306 Coloured Plates, bound in half morocco. gilt edges, £11 11s.

CHEMISTRY.

BERNAYS.—Notes for Students in Chemistry;
being a Syllabus of Chemistry compiled mainly from the Manuals of Fownes-Watts, Miller, Wurz, and Schorlemmer. By ALBERT J. BERNAYS, Ph.D., Professor of Chemistry at St. Thomas's Hospital. Sixth Edition. Fcap. 8vo, 3s. 6d.

By the same Author.

Skeleton· Notes on Analytical Chemistry,
for Students in Medicine. Fcap. 8vo, 2s. 6d.

BLOXAM.—Chemistry, Inorganic and Organic;
with Experiments. By CHARLES L. BLOXAM, Professor of Chemistry in King's College. Fifth Edition. With 292 Engravings. 8vo, 16s.

By the same Author.

Laboratory Teaching; or, Progressive
Exercises in Practical Chemistry. Fourth Edition. With 83 Engravings. Crown 8vo, 5s. 6d.

11, *NEW BURLINGTON STREET.*

J. & A. Churchill's Medical Class Books.

CHEMISTRY—*continued*.

BOWMAN AND BLOXAM.—**Practical Chemistry,**
including Analysis. By JOHN E. BOWMAN, formerly Professor of Practical Chemistry in King's College, and CHARLES L. BLOXAM, Professor of Chemistry in King's College. With 98 Engravings. Seventh Edition. Fcap. 8vo, 6s. 6d.

BROWN. — **Practical Chemistry: Analytical**
Tables and Exercises for Students. By J. CAMPBELL BROWN, D.Sc. Lond., Professor of Chemistry in University College, Liverpool. Second Edition. 8vo, 2s. 6d.

CLOWES.—**Practical Chemistry and Qualita**-tive Inorganic Analysis. An Elementary Treatise, specially adapted for use in the Laboratories of Schools and Colleges, and by Beginners. By FRANK CLOWES, D.Sc., Professor of Chemistry in University College, Nottingham. Third Edition. With 47 Engravings. Post 8vo, 7s. 6d.

FOWNES.—**Manual of Chemistry.**—*See WATTS.*

FRANKLAND AND JAPP.—**Inorganic Chemistry.**
By EDWARD FRANKLAND, Ph.D., D.C.L., F.R.S., and F. R. JAPP, M.A., Ph.D., F.I.C. With 2 Lithographic Plates and numerous Wood Engravings. 8vo, 24s.

LUFF.—**An Introduction to the Study of Che**-mistry. Specially designed for Medical and Pharmaceutical Students. By A. P. LUFF, F.I.C., F.C.S., Lecturer on Chemistry in the Central School of Chemistry and Pharmacy. Crown 8vo, 2s. 6d.

TIDY.—**A Handbook of Modern Chemistry,**
Inorganic and Organic. By C. MEYMOTT TIDY, M.B., Professor of Chemistry and Medical Jurisprudence at the London Hospital, 8vo, 16s.

VACHER.—**A Primer of Chemistry, including**
Analysis. By ARTHUR VACHER. 18mo, 1s.

VALENTIN.—**Chemical Tables for the Lecture**-room and Laboratory. By WILLIAM G. VALENTIN, F.C.S. In Five large Sheets, 5s. 6d.

11, *NEW BURLINGTON STREET.*

J. & A. Churchill's Medical Class Books.

CHEMISTRY—*continued.*

VALENTIN AND HODGKINSON.—A Course of
Qualitative Chemical Analysis. By W. G. VALENTIN, F.C.S. Sixth Edition by W. R. HODGKINSON, Ph.D. (Wurzburg), Senior Demonstrator of Practical Chemistry in the Science Schools, South Kensington, and H. M. CHAPMAN, Assistant Demonstrator. With Engravings. 8vo, 8s. 6d.

WATTS.—Physical and Inorganic Chemistry.
By HENRY WATTS, B.A., F.R.S. (being Vol. I. of the Thirteenth Edition of Fownes' Manual of Chemistry). With 150 Wood Engravings, and Coloured Plate of Spectra. Crown 8vo, 9s.

By the same Author.

Chemistry of Carbon-Compounds, or
Organic Chemistry (being Vol. II. of the Twelfth Edition of Fownes' Manual of Chemistry). With Engravings. Crown 8vo, 10s.

CHILDREN, DISEASES OF.

DAY.—A Treatise on the Diseases of Children.
For Practitioners and Students. By WILLIAM H. DAY, M.D., Physician to the Samaritan Hospital for Women and Children. Crown 8vo, 12s. 6d.

ELLIS.—A Practical Manual of the Diseases
of Children. By EDWARD ELLIS, M.D., late Senior Physician to the Victoria Hospital for Sick Children. With a Formulary. Fourth Edition. Crown 8vo, 10s.

SMITH.—On the Wasting Diseases of Infants
and Children. By EUSTACE SMITH, M.D., F.R.C.P., Physician to H.M. the King of the Belgians, and to the East London Hospital for Children. Fourth Edition. Post 8vo, 8s. 6d.

By the same Author.

A Practical Treatise on Disease in Chil-
dren. 8vo, 22s.

STEINER.—Compendium of Children's Dis-
eases; a Handbook for Practitioners and Students. By JOHANN STEINER, M.D. Translated by LAWSON TAIT, F.R.C.S., Surgeon to the Birmingham Hospital for Women, &c. 8vo, 12s. 6d.

11, *NEW BURLINGTON STREET.*

J. & A. Churchill's Medical Class Books.

DENTISTRY.

GORGAS.—Dental Medicine: a Manual of
Dental Materia Medica and Therapeutics, for Practitioners and Students. By FERDINAND J. S. GORGAS, A.M., M.D., D.D.S., Professor of Dentistry in the University of Maryland; Editor of "Harris's Principles and Practice of Dentistry," &c. Royal 8vo, 14s.

SEWILL.—The Student's Guide to Dental
Anatomy and Surgery. By HENRY E. SEWILL, M.R.C.S., L.D.S., late Dental Surgeon to the West London Hospital. Second Edition. With 78 Engravings. Fcap. 8vo, 5s. 6d.

STOCKEN.—Elements of Dental Materia Medica
and Therapeutics, with Pharmacopœia. By JAMES STOCKEN, L.D.S.R.C.S., late Lecturer on Dental Materia Medica and Therapeutics and Dental Surgeon to the National Dental Hospital; assisted by THOMAS GADDES, L.D.S. Eng. and Edin. Third Edition. Fcap. 8vo, 7s. 6d.

TAFT.—A Practical Treatise on Operative
Dentistry. By JONATHAN TAFT, D.D.S., Professor of Operative Surgery in the Ohio College of Dental Surgery. Third Edition. With 134 Engravings. 8vo, 18s.

TOMES (C. S.).—Manual of Dental Anatomy,
Human and Comparative. By CHARLES S. TOMES, M.A., F.R.S. Second Edition. With 191 Engravings. Crown 8vo, 12s. 6d.

TOMES (J. and C. S.).—A Manual of Dental
Surgery. By JOHN TOMES, M.R.C.S., F.R.S., and CHARLES S. TOMES, M.A., M.R.C.S., F.R.S.; Lecturer on Anatomy and Physiology at the Dental Hospital of London. Third Edition. With many Engravings, Crown 8vo. [*In the press.*

EAR, DISEASES OF.

BURNETT.—The Ear: its Anatomy, Physio-
logy, and Diseases. A Practical Treatise for the Use of Medical Students and Practitioners. By CHARLES H. BURNETT, M.D., Aural Surgeon to the Presbyterian Hospital, Philadelphia. With 87 Engravings. 8vo, 18s.

DALBY.—On Diseases and Injuries of the Ear.
By WILLIAM B. DALBY, F.R.C.S., Aural Surgeon to, and Lecturer on Aural Surgery at, St. George's Hospital. Second Edition. With Engravings. Fcap. 8vo, 6s. 6d.

11, *NEW BURLINGTON STREET.*

EAR, DISEASES OF—continued.

JONES.—A Practical Treatise on Aural Surgery. By H. MACNAUGHTON JONES, M.D., Professor of the Queen's University in Ireland, late Surgeon to the Cork Ophthalmic and Aural Hospital. Second Edition. With 63 Engravings. Crown 8vo, 8s. 6d.

By the same Author.

Atlas of the Diseases of the Membrana Tympani. In Coloured Plates, containing 59 Figures. With Explanatory Text. Crown 4to, 21s.

FORENSIC MEDICINE.

OGSTON.—Lectures on Medical Jurisprudence. By FRANCIS OGSTON, M.D., late Professor of Medical Jurisprudence and Medical Logic in the University of Aberdeen. Edited by FRANCIS OGSTON, Jun., M.D., late Lecturer on Practical Toxicology in the University of Aberdeen. With 12 Plates. 8vo, 18s.

TAYLOR.—The Principles and Practice of Medical Jurisprudence. By ALFRED S. TAYLOR, M.D., F.R.S. Third Edition, revised by THOMAS STEVENSON, M.D., F.R.C.P., Lecturer on Chemistry and Medical Jurisprudence at Guy's Hospital; Examiner in Chemistry at the Royal College of Physicians; Official Analyst to the Home Office. With 188 Engravings. 2 Vols. 8vo, 31s. 6d.

By the same Author.

A Manual of Medical Jurisprudence. Tenth Edition. With 55 Engravings. Crown 8vo, 14s.

ALSO,

On Poisons, in relation to Medical Jurisprudence and Medicine. Third Edition. With 104 Engravings. Crown 8vo, 16s.

TIDY AND WOODMAN.—A Handy-Book of Forensic Medicine and Toxicology. By C. MEYMOTT TIDY, M.B.; and W. BATHURST WOODMAN, M.D., F.R.C.P. With 8 Lithographic Plates and 116 Wood Engravings. 8vo, 31s. 6d.

J. & A. Churchill's Medical Class Books.

HYGIENE.

PARKES.—A Manual of Practical Hygiene.
By EDMUND A. PARKES, M.D., F.R.S. Sixth Edition by F. DE CHAUMONT, M.D., F.R.S., Professor of Military Hygiene in the Army Medical School. With 9 Plates and 103 Engravings. 8vo, 18s.

WILSON.—A Handbook of Hygiene and Sanitary Science. By GEORGE WILSON, M.A., M.D., F.R.S.E., Medical Officer of Health for Mid Warwickshire. Fifth Edition. With Engravings. Crown 8vo, 10s. 6d.

MATERIA MEDICA AND THERAPEUTICS.

BINZ AND SPARKS.—The Elements of Therapeutics; a Clinical Guide to the Action of Medicines. By C. BINZ, M.D., Professor of Pharmacology in the University of Bonn. Translated and Edited with Additions, in conformity with the British and American Pharmacopœias, by EDWARD I. SPARKS, M.A., M.B., F.R.C.P. Lond. Crown 8vo, 8s. 6d.

OWEN.—A Manual of Materia Medica; incorporating the Author's "Tables of Materia Medica." By ISAMBARD OWEN, M.D., Lecturer on Materia Medica and Therapeutics to St. George's Hospital. Crown 8vo, 6s.

ROYLE AND HARLEY.—A Manual of Materia Medica and Therapeutics. By J. FORBES ROYLE, M.D., F.R.S., and JOHN HARLEY, M.D., F.R.C.P., Physician to, and Joint Lecturer on Clinical Medicine at, St. Thomas's Hospital. Sixth Edition. With 139 Engravings. Crown 8vo, 15s.

THOROWGOOD.—The Student's Guide to Materia Medica and Therapeutics. By JOHN C. THOROWGOOD, M.D., F.R.C.P., Lecturer on Materia Medica at the Middlesex Hospital. Second Edition. With Engravings. Fcap. 8vo, 7s.

WARING.—A Manual of Practical Therapeutics. By EDWARD J. WARING, C.I.E., M.D., F.R.C.P. Third Edition. Fcap. 8vo, 12s. 6d.

11, NEW BURLINGTON STREET.

J. & A. Churchill's Medical Class Books.

MEDICINE.

BARCLAY.—A Manual of Medical Diagnosis.
By A. WHYTE BARCLAY, M.D., F.R.C.P., late Physician to, and Lecturer on Medicine at, St. George's Hospital. Third Edition. Fcap. 8vo, 10s. 6d.

CHARTERIS.—The Student's Guide to the Practice of Medicine. By MATTHEW CHARTERIS, M.D., Professor of Materia Medica, University of Glasgow; Physician to the Royal Infirmary. With Engravings on Copper and Wood. Third Edition. Fcap. 8vo, 7s.

FENWICK.—The Student's Guide to Medical Diagnosis. By SAMUEL FENWICK, M.D., F.R.C.P., Physician to the London Hospital. Fifth Edition. With 111 Engravings. Fcap. 8vo, 7s.

By the same Author.
The Student's Outlines of Medical Treatment. Second Edition. Fcap. 8vo, 7s.

FLINT.—Clinical Medicine: a Systematic Treatise on the Diagnosis and Treatment of Disease. By AUSTIN FLINT, M.D., Professor of the Principles and Practice of Medicine, &c., in Bellevue Hospital Medical College. 8vo, 20s.

HALL.—Synopsis of the Diseases of the Larynx, Lungs, and Heart: comprising Dr. Edwards' Tables on the Examination of the Chest. With Alterations and Additions. By F. DE HAVILLAND HALL, M.D., F.R.C.P., Assistant-Physician to the Westminster Hospital. Royal 8vo, 2s. 6d.

SANSOM.—Manual of the Physical Diagnosis of Diseases of the Heart, including the use of the Sphygmograph and Cardiograph. By A. E. SANSOM, M.D., F.R.C.P., Assistant-Physician to the London Hospital. Third Edition. With 47 Woodcuts. Fcap. 8vo, 7s. 6d.

WARNER.—Student's Guide to Clinical Medicine and Case-Taking. By FRANCIS WARNER, M.D., F.R.C.P., Assistant-Physician to the London Hospital. Second Edition. Fcap. 8vo, 5s.

WEST.—How to Examine the Chest: being a Practical Guide for the Use of Students. By SAMUEL WEST, M.D., M.R.C.P., Physician to the City of London Hospital for Diseases of the Chest, &c. With 42 Engravings. Fcap. 8vo, 5s.

J. & A. Churchill's Medical Class Books.

MEDICINE—*continued.*

WHITTAKER.—Student's Primer on the Urine.
By J. TRAVIS WHITTAKER, M.D., Clinical Demonstrator at the Royal Infirmary, Glasgow. With Illustrations, and 16 Plates etched on Copper. Post 8vo, 4s. 6d.

MIDWIFERY.

BARNES.—Lectures on Obstetric Operations,
including the Treatment of Hæmorrhage, and forming a Guide to the Management of Difficult Labour. By ROBERT BARNES, M.D., F.R.C.P., Obstetric Physician to, and Lecturer on Diseases of Women, &c., at, St. George's Hospital. Third Edition. With 124 Engravings. 8vo, 18s.

CLAY.—The Complete Handbook of Obstetric
Surgery; or, Short Rules of Practice in every Emergency, from the Simplest to the most formidable Operations connected with the Science of Obstetricy. By CHARLES CLAY, M.D., late Senior Surgeon to, and Lecturer on Midwifery at, St. Mary's Hospital, Manchester. Third Edition. With 91 Engravings. Fcap. 8vo, 6s. 6d.

RAMSBOTHAM.—The Principles and Practice
of Obstetric Medicine and Surgery. By FRANCIS H. RAMSBOTHAM, M.D., formerly Obstetric Physician to the London Hospital. Fifth Edition. With 120 Plates, forming one thick handsome volume. 8vo, 22s.

REYNOLDS. — Notes on Midwifery: specially
designed to assist the Student in preparing for Examination. By J. J. REYNOLDS, L.R.C.P., M.R.C.S. Fcap. 8vo, 4s.

ROBERTS.—The Student's Guide to the Practice
of Midwifery. By D. LLOYD ROBERTS, M.D., F.R.C.P., Physician to St. Mary's Hospital, Manchester. Third Edition. With 2 Coloured Plates and 127 Engravings. Fcap. 8vo, 7s. 6d.

SCHROEDER.—A Manual of Midwifery; includ-
ing the Pathology of Pregnancy and the Puerperal State. By KARL SCHROEDER, M.D., Professor of Midwifery in the University of Erlangen. Translated by CHARLES H. CARTER, M.D. With Engravings. 8vo, 12s. 6d.

SWAYNE.—Obstetric Aphorisms for the Use of
Students commencing Midwifery Practice. By JOSEPH G. SWAYNE, M.D., Lecturer on Midwifery at the Bristol School of Medicine. Eighth Edition. With Engravings. Fcap. 8vo, 3s. 6d.

11, *NEW BURLINGTON STREET.*

J. & A. Churchill's Medical Class Books.

MICROSCOPY.

CARPENTER.—**The Microscope and its Revela**tions. By WILLIAM B. CARPENTER, C.B., M.D., F.R.S. Sixth Edition. With 26 Plates, a Coloured Frontispiece, and more than 500 Engravings. Crown 8vo, 16s.

MARSH. — **Microscopical Section-Cutting : a** Practical Guide to the Preparation and Mounting of Sections for the Microscope, special prominence being given to the subject of Animal Sections. By Dr. SYLVESTER MARSH. Second Edition. With 17 Engravings. Fcap. 8vo, 3s. 6d.

MARTIN.—**A Manual of Microscopic Mounting.** By JOHN H. MARTIN, Member of the Society of Public Analysis, &c. Second Edition. With several Plates and 144 Engravings. 8vo, 7s. 6d.

OPHTHALMOLOGY.

HIGGENS.—**Hints on Ophthalmic Out-Patient** Practice. By CHARLES HIGGENS, F.R.C.S., Ophthalmic Surgeon to, and Lecturer on Ophthalmology at, Guy's Hospital. Second Edition. Fcap. 8vo, 3s.

JONES.—**A Manual of the Principles and** Practice of Ophthalmic Medicine and Surgery. By T. WHARTON JONES, F.R.C.S., F.R.S., late Ophthalmic Surgeon and Professor of Ophthalmology to University College Hospital. Third Edition. With 9 Coloured Plates and 173 Engravings. Fcap. 8vo, 12s. 6d.

NETTLESHIP.—**The Student's Guide to Diseases** of the Eye. By EDWARD NETTLESHIP, F.R.C.S., Ophthalmic Surgeon to, and Lecturer on Ophthalmic Surgery at, St. Thomas's Hospital. Third Edition. With 157 Engravings, and a Set of Coloured Papers illustrating Colour-blindness. Fcap. 8vo, 7s. 6d.

TOSSWILL.—**Diseases and Injuries of the Eye** and Eyelids. By LOUIS H. TOSSWILL, B.A., M.B. Cantab., M.R.C.S., Surgeon to the West of England Eye Infirmary, Exeter. Fcap. 8vo, 2s. 6d.

WOLFE.—**On Diseases and Injuries of the Eye :** a Course of Systematic and Clinical Lectures to Students and Medical Practitioners. By J. R. WOLFE, M.D., F.R.C.S.E., Senior Surgeon to the Glasgow Ophthalmic Institution, Lecturer on Ophthalmic Medicine and Surgery in Anderson's College. With 10 Coloured Plates, and 120 Wood Engravings, 8vo, 21s.

11, NEW BURLINGTON STREET.

J. & A. Churchill's Medical Class Books.

PATHOLOGY.

JONES AND SIEVEKING.—A Manual of Pathological Anatomy. By C. HANDFIELD JONES, M.B., F.R.S., and EDWARD H. SIEVEKING, M.D., F.R.C.P. Second Edition. Edited, with considerable enlargement, by J. F. PAYNE, M.B., Assistant-Physician and ecturer on General Pathology at St. Thomas's Hospital. With 195 Engravings. Crown 8vo, 16s.

LANCEREAUX.—Atlas of Pathological Anatomy. By Dr. LANCEREAUX. Translated by W. S. GREENFIELD, M.D., Professor of Pathology in the University of Edinburgh. With 70 Coloured Plates. Imperial 8vo, £5 5s.

VIRCHOW. — Post-Mortem Examinations: a Description and Explanation of the Method of Performing them, with especial reference to Medico-Legal Practice. By Professor RUDOLPH VIRCHOW, Berlin Charité Hospital. Translated by Dr. T. B. SMITH. Second Edition, with 4 Plates. Fcap. 8vo, 3s. 6d.

WILKS AND MOXON.—Lectures on Pathological Anatomy. By SAMUEL WILKS, M.D., F.R.S., Physician to, and late Lecturer on Medicine at, Guy's Hospital; and WALTER MOXON, M.D., F.R.C.P., Physician to, and Lecturer on the Practice of Medicine at, Guy's Hospital. Second Edition. With 7 Steel Plates. 8vo, 18s.

PSYCHOLOGY.

BUCKNILL AND TUKE.—A Manual of Psychological Medicine: containing the Lunacy Laws, Nosology, Ætiology, Statistics, Description, Diagnosis, Pathology, and Treatment of Insanity, with an Appendix of Cases. By JOHN C. BUCKNILL, M.D., F.R.S., and D. HACK TUKE, M.D., F.R.C.P. Fourth Edition with 12 Plates (30 Figures). 8vo, 25s.

CLOUSTON. — Clinical Lectures on Mental Diseases. By THOMAS S. CLOUSTON, M.D., and F.R.C.P. Edin.; Lecturer on Mental Diseases in the University of Edinburgh. With 8 Plates (6 Coloured). Crown 8vo, 12s. 6d.

MANN.—A Manual of Psychological Medicine and Allied Nervous Disorders. By EDWARD C. MANN, M.D., Member of the New York Medico-Legal Society. With Plates. 8vo, 24s.

11, *NEW BURLINGTON STREET.*

J. & A. Churchill's Medical Class Books.

PHYSIOLOGY.

CARPENTER.—Principles of Human Physiology. By WILLIAM B. CARPENTER, C.B., M.D., F.R.S. Ninth Edition. Edited by Henry Power, M.B., F.R.C.S. With 3 Steel Plates and 377 Wood Engravings. 8vo, 31s. 6d.

DALTON.—A Treatise on Human Physiology: designed for the use of Students and Practitioners of Medicine. By JOHN C. DALTON, M.D., Professor of Physiology and Hygiene in the College of Physicians and Surgeons, New York. Seventh Edition. With 252 Engravings. Royal 8vo, 20s.

FREY.—The Histology and Histo-Chemistry of Man. A Treatise on the Elements of Composition and Structure of the Human Body. By HEINRICH FREY, Professor of Medicine in Zurich. Translated by ARTHUR E. BARKER, Assistant-Surgeon to the University College Hospital. With 608 Engravings. 8vo, 21s.

SANDERSON.—Handbook for the Physiological Laboratory: containing an Exposition of the fundamental facts of the Science, with explicit Directions for their demonstration. By J. BURDON SANDERSON, M.D., F.R.S.; E. KLEIN, M.D., F.R.S.; MICHAEL FOSTER, M.D., F.R.S.; and T. LAUDER BRUNTON, M.D., F.R.S. 2 Vols., with 123 Plates. 8vo, 24s.

YEO.—A Manual of Physiology for the Use of Junior Students of Medicine. By GERALD F. YEO, M.D., F.R.C.S., Professor of Physiology in King's College, London. With 301 Engravings. Crown 8vo, 14s.

SURGERY.

BELLAMY.—The Student's Guide to Surgical Anatomy; a Description of the more important Surgical Regions of the Human Body, and an Introduction to Operative Surgery. By EDWARD BELLAMY, F.R.C.S., and Member of the Board of Examiners; Surgeon to, and Lecturer on Anatomy at, Charing Cross Hospital. Second Edition. With 76 Engravings. Fcap. 8vo, 7s.

BRYANT.—A Manual for the Practice of Surgery. By THOMAS BRYANT, F.R.C.S., Surgeon to, and Lecturer on Surgery at, Guy's Hospital. Fourth Edition. With 750 Illustrations (many being coloured), and including 6 Chromo-Lithographic Plates. 2 Vols. Crown 8vo, 32s.

J. & A. Churchill's Medical Class Books.

SURGERY—continued.

CLARK AND WAGSTAFFE. — Outlines of Surgery and Surgical Pathology. By F. LE GROS CLARK, F.R.C.S., F.R.S., Consulting Surgeon to St. Thomas's Hospital. Second Edition. Revised and expanded by the Author, assisted by W. W. WAGSTAFFE, F.R.C.S., Assistant Surgeon to St. Thomas's Hospital. 8vo, 10s. 6d.

DRUITT. — The Surgeon's Vade-Mecum; a Manual of Modern Surgery. By ROBERT DRUITT, F.R.C.S. Eleventh Edition. With 369 Engravings. Fcap. 8vo, 14s.

FERGUSSON. — A System of Practical Surgery. By Sir WILLIAM FERGUSSON, Bart., F.R.C.S., F.R.S., late Surgeon and Professor of Clinical Surgery to King's College Hospital. With 463 Engravings. Fifth Edition. 8vo, 21s.

HEATH. — A Manual of Minor Surgery and Bandaging, for the use of House-Surgeons, Dressers, and Junior Practitioners. By CHRISTOPHER HEATH, F.R.C.S., Holme Professor of Clinical Surgery in University College and Surgeon to the Hospital. Seventh Edition. With 129 Engravings. Fcap. 8vo, 6s.

By the same Author.

A Course of Operative Surgery: with Twenty Plates (containing many figures) drawn from Nature by M. LÉVEILLÉ, and Coloured. Second Edition. Large 8vo, 30s.

ALSO,

The Student's Guide to Surgical Diagnosis. Second Edition. Fcap. 8vo, 6s. 6d.

MAUNDER. — Operative Surgery. By Charles F. MAUNDER, F.R.C.S., late Surgeon to, and Lecturer on Surgery at, the London Hospital. Second Edition. With 164 Engravings. Post 8vo, 6s.

SOUTHAM. — Regional Surgery: including Surgical Diagnosis. A Manual for the use of Students. BY FREDERICK A. SOUTHAM, M.A., M.B. Oxon, F.R.C.S., Assistant-Surgeon to the Royal Infirmary, and Assistant-Lecturer on Surgery in the Owen's College School of Medicine, Manchester.
  Part I. The Head and Neck. Crown 8vo, 6s. 6d.
  „ II. The Upper Extremity and Thorax. Crown 8vo, 7s. 6d.

J. & A. Churchill's Medical Class Books.

TERMINOLOGY.

DUNGLISON.—Medical Lexicon : a Dictionary
of Medical Science, containing a concise Explanation of its various Subjects and Terms, with Accentuation, Etymology, Synonyms, &c. By ROBERT DUNGLISON, M.D. New Edition, thoroughly revised by RICHARD J. DUNGLISON, M.D. Royal 8vo, 28s.

MAYNE.—A Medical Vocabulary : being an
Explanation of all Terms and Phrases used in the various Departments of Medical Science and Practice, giving their Derivation, Meaning, Application, and Pronunciation. By ROBERT G. MAYNE, M.D., LL.D., and JOHN MAYNE, M.D., L.R.C.S.E. Fifth Edition. Crown 8vo, 10s. 6d.

WOMEN, DISEASES OF.

BARNES.—A Clinical History of the Medical
and Surgical Diseases of Women. By ROBERT BARNES, M.D., F.R.C.P., Obstetric Physician to, and Lecturer on Diseases of Women, &c., at, St. George's Hospital. Second Edition. With 181 Engravings. 8vo, 28s.

COURTY.—Practical Treatise on Diseases of
the Uterus, Ovaries, and Fallopian Tubes. By Professor COURTY, Montpellier. Translated from the Third Edition by his Pupil, AGNES M'LAREN, M.D., M.K.Q.C.P. With Preface by Dr. MATTHEWS DUNCAN. With 424 Engravings. 8vo, 24s.

DUNCAN.—Clinical Lectures on the Diseases
of Women. By J. MATTHEWS DUNCAN, M.D., F.R.C.P., F.R.S.E., Obstetric Physician to St. Bartholomew's Hospital. Second Edition, with Appendices. 8vo, 14s.

EMMET. — The Principles and Practice of
Gynæcology. By THOMAS ADDIS EMMET, M.D., Surgeon to the Woman's Hospital of the State of New York. With 130 Engravings. Royal 8vo, 24s.

GALABIN.—The Student's Guide to the Diseases of Women. By ALFRED L. GALABIN, M.D., F.R.C.P., Obstetric Physician to, and Lecturer on Obstetric Medicine at, Guy's Hospital. Third Edition. With 78 Engravings. Fcap. 8vo, 7s. 6d.

11, *NEW BURLINGTON STREET.*

WOMEN, DISEASES OF—continued.

REYNOLDS.—Notes on Diseases of Women.
Specially designed to assist the Student in preparing for Examination. By J. J. REYNOLDS, L.R.C.P., M.R.C.S. Second Edition. Fcap. 8vo 2s. 6d.

SAVAGE.—The Surgery of the Female Pelvic
Organs. By HENRY SAVAGE, M.D., Lond., F.R.C.S., one of the Consulting Medical Officers of the Samaritan Hospital for Women. Fifth Edition, with 17 Lithographic Plates (15 Coloured), and 52 Woodcuts. Royal 4to, 35s.

SMITH.—Practical Gynæcology: a Handbook
of the Diseases of Women. By HEYWOOD SMITH, M.D., Physician to the Hospital for Women and to the British Lying-in Hospital. With Engravings. Second Edition. Crown 8vo. [*In preparation*.

WEST AND DUNCAN.—Lectures on the Diseases of Women. By CHARLES WEST, M.D., F.R.C.P. Fourth Edition. Revised and in part re-written by the Author, with numerous additions by J. MATTHEWS DUNCAN, M.D., F.R.C.P., F.R.S.E., Obstetric Physician to St. Bartholomew's Hospital. 8vo, 16s.

ZOOLOGY.

CHAUVEAU AND FLEMING.—The Comparative Anatomy of the Domesticated Animals. By A. CHAUVEAU, Professor at the Lyons Veterinary School; and GEORGE FLEMING, Veterinary Surgeon, Royal Engineers. With 450 Engravings. 8vo, 31s. 6d.

HUXLEY.—Manual of the Anatomy of Invertebrated Animals. By THOMAS H. HUXLEY, LL.D., F.R.S. With 156 Engravings. Post 8vo, 16s.

By the same Author.

Manual of the Anatomy of Vertebrated
Animals. With 110 Engravings. Post 8vo, 12s.

WILSON.—The Student's Guide to Zoology:
a Manual of the Principles of Zoological Science. By ANDREW WILSON, Lecturer on Natural History, Edinburgh. With Engravings. Fcap. 8vo, 6s. 6d.

11, *NEW BURLINGTON STREET*.

www.ingramcontent.com/pod-product-compliance
Lightning Source LLC
Chambersburg PA
CBHW050847300426
44111CB00010B/1169